# Introduction to
# Field Biology

## Second Edition

**DONALD P. BENNETT, M.A., F.I.Biol.**
*Head of the Biology Department*
*Cheltenham Grammar School*

*and*

**DAVID A. HUMPHRIES, B.Sc., Ph.D., M.I.Biol.**
*University of Aston in Birmingham*

EDWARD ARNOLD

© Donald P. Bennett and David A. Humphries 1974

First published 1965
by Edward Arnold (Publishers) Ltd.
25 Hill Street, London WIX 8LL
Reprinted 1967, 1970
Second edition 1974
Reprinted 1976
Reprinted 1977
Reprinted with amendments 1979

Boards edition ISBN: 0 7131 2459 8
Paper edition ISBN: 0 7131 2458 X

*Printed and bound in Great Britain by*
*William Clowes (Beccles) Limited, Beccles and London*

# Preface to the Second Edition

Since this book was first published, there have been significant and welcome changes in the amount of attention paid by mass media to matters concerning the environment, so that the terms *ecology* and *conservation* have become familiar even if they are often incompletely understood. Ecological disasters, such as the massive oil pollution from wrecked tankers and the even more damaging use of detergent to disperse the mess, have emphasized the need to understand how changes in the environment can affect living organisms. For it is only through fuller understanding that the most desirable relationships between human activities and animals, plants, and their habitats, can be conserved. We hope this revised text will continue to play a useful part in advancing this understanding.

In revising the book, we have introduced the concepts of energy-flow, productivity, and population dynamics, and have devoted more attention to the problems of pollution. There is now a section estuaries, a new chapter on Town Ecology and an App ... outlining some methods for investigating organic pollution; the r ading list has also been brought up to date and expanded. To help the individual worker, considerable additional material on project work has been included, for which our experiences at Cheltenham Grammar School and at the University of Aston have provided many new suggestions.

This book was written to meet the need for an introductory text which attempted from the start to show how fieldwork techniques should be used to clarify and solve problems concerning both animals and plants. Whilst it is primarily intended for students taking biology courses in sixth forms, Colleges of Education or Universities, experience shows that it has also proved useful to students following Liberal Studies courses.

## Acknowledgements

Many of the results quoted and all the half-tone plates were obtained during field activities of individuals or groups from Cheltenham Grammar School during the last twenty years. The school has a long tradition of field studies and owes much to F. Craven Broad, E. J. Machin, R. E. Lister and Dr. P. M. Driver for the parts they played in helping to create this. Help, advice and information have come from many quarters and it is a special pleasure to acknowledge the assistance of the following: M. J. Cotton, H. J. Fletcher, M. D. Hayward, M. F.

Heyworth, N. B. Palmer, R. D. Ransome, (all past members of C. G. S. Biology Society), D. M. Barling, H. F. Howard, T. P. Hardie, and D. B. Lewis. The new material on pollution and urban ecology owes much to discussions with Dr. D. Allsopp and Mr. H. A. Hawkes of the Department of Biological Sciences, Aston University. Some of the art work was undertaken by Eric Ford when he was Art master at C. G. S. Grateful acknowledgement is given for permission to reproduce the following: Fig. 6.1 (Prof. E. B. Edney), Fig. 6.13 (Prof. E. B. Ford), Fig. 9.6 (Prof. J. L. Harper), Fig. 10.8 (Dr. H. B. Cott), Fig. 14.2 (Dr. O. L. Gilbert). We are indebted to the Literary Executory of the late Sir Ronald A. Fisher, F.R.S., Cambridge, to Dr. Frank Yates, F.R.S., Rothamsted, and to Oliver & Boyd Ltd., Edinburgh, for permission to include in Table 5.3 figures from their book *Statistical Tables for Biological, Agricultural and Medical Research*. Much of the material for the present revision was assembled by D. P. B. during his tenure of the Ward-Perkins Studentship at Pembroke College, Oxford, and he gratefully acknowledges his debt to the college.

Cheltenham　　　　　　　　　　　　　　　　　　　　　　　D.P.B.
1974　　　　　　　　　　　　　　　　　　　　　　　　　　　D.A.H.

# From the Preface to the First Edition

At a time when progress in science is increasingly dependent on expensive and complicated apparatus, Field Biology remains practically the only branch of the subject in which it is still possible for the beginner to make important original discoveries. Nearly every field survey will provide problems of distribution and adaptation which may be solved readily by observation or experiment. The problems are never quite the same from one area to the next because so many different and variable factors are involved. The intellectual challenge provided by fieldwork is now widely recognized and all recent proposals for changes in school biology syllabuses have stressed the importance of field investigations because experience shows that the scientific approach of critical but objective enquiry and open-mindedness soon begins to infect classroom and laboratory work as well.

In the space available it has not been possible to give more than an introductory sketch of the major habitats, and to consider for each one or two topics in more detail. Thus various habitat factors, competition, stratification, adaptation, zonation, and succession have all been considered for at least one and often more than one habitat. In general, descriptive accounts have been given only where they illustrate particular relationships between organism and environment. Questions are asked in the hope that some readers will seek and find the answers appropriate to their own locality.

# Contents

# 1 Methods of Discovery

Have you ever wondered why animals or plants found in one area are rarely found in apparently similar places elsewhere? Why, for instance, are foxgloves common in hedgerows around Dartmoor but usually absent from those in the Cotswolds? And why is the crayfish common in Cotswold streams but not in the Dartmoor rivers? Is there any link between foxglove and crayfish distribution? If you have indeed asked yourself this kind of question, you are well on the way to becoming a Field Biologist.

The main objects of this book are to help you to approach such problems scientifically and to aid you in acquiring practical experience in the many techniques involved. The 'Scientific Approach' which we must employ consists in planning systematically a line of enquiry, and checking and cross-checking at every stage in the proceedings. It is essential to ask the right questions, so our first move in dealing with the foxgloves and crayfish must be to check the original observations by making a detailed and systematic survey. Foxgloves are most conspicuous when in flower – perhaps we missed seeing them in the Cotswolds because they were in a vegetative state. Crayfish are elusive creatures at the best of times, so we must make a thorough search in our Dartmoor rivers before deciding that they really are absent.

When we are certain our observations are correct and that we did ask the right questions, we are ready to investigate further. What we try to do is to make an intelligent guess at the answer, and then test it by experiment. We may have read that the Cotswolds are based on calcareous rock formations. So we might guess that absence of lime on Dartmoor may be responsible for both the presence of foxgloves and the lack of crayfish.

We could test the effect of lime on foxgloves by adding it to soil in which they are growing, and then comparing their progress with foxgloves growing alongside in untreated soil. It would be more difficult to modify conditions in the field for active animals like crayfish, so in their case we might observe them in similar aquarium tanks, one containing 'hard' water from the Cotswolds and the other containing lime-deficient water. If we have guessed correctly, these experiments should give us the evidence we need: even if they do not give the expected answer, they are making a positive addition to our knowledge. In either case, we must be patient because living things may take weeks or even months to respond to environmental changes of this kind.

The methods outlined so far form part of the general pattern of all serious biological studies. In general we may say that all such studies begin with *observations* which arouse enough interest to stimulate *description*. Because it is impossible to describe everything in detail, we select only the aspects which interest us. In our example, these were the distribution of foxgloves and crayfish in two different areas. After a preliminary look, we break down our chosen aspect into its 'elements'. These are the basic facts that we wish to record. If we are recording animal and plant distribution, our basic facts will be the different organisms we have identified. It is obvious that we must learn to identify accurately, and this is where practice and experience count – a generalization about 'buttercups' or 'earthworms' could be based on half-a-dozen instead of a single species. We shall also need notes, maps and sketches to record when and where we saw them,

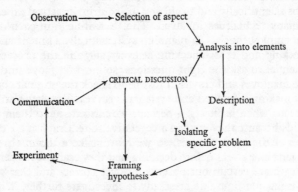

**Fig. 1.1**   An outline of scientific methods

what they were doing and what they looked like. When, as sometimes we set out to describe a bird's behaviour pattern, it is useful to decide first on a list of basic activities such as 'fly', 'alight', 'head on one side', 'peck' and so on, and for speed in jotting them down a shorthand version might be used. As information is collected, we begin to classify it in ways that are likely to be useful. Thus in woodland we might list all the plants according to their life forms (trees, shrubs, herbs, mosses) or by their families. Animals might be grouped by feeding habits or by habitats such as bark and leaf-litter.

A very important part of the descriptive work will be the measuring and recording of *habitat factors* which may affect the distribution and behaviour of the animals and plants (see Chapters 6 and 7). Much of the first half of this book is concerned with making scientific records because without solid facts we cannot make useful discoveries. While clear records are essential, they must not be regarded as ends in themselves but rather as means of discovering what the problems are, and how to solve them. Later in the book there are some suggestions about topics worth investigating in most major habitats. There are also a number of questions which can be answered by careful observation and experiment, but it is much more valuable for you to find and answer problems of your own. One useful approach is to select any relationship which interests you and look for *cause* and *effect*. Thus we ask 'Why?' or 'What causes this?', and then 'What effect does this have on other things?' We might ask why a certain tree bears leaves only in the summer months. Our answer should provide an explanation, possibly in terms of the tree's requirements for transpiration and photosynthesis. But our approach is not complete unless we also ask how this annual cycle of leaf growth affects other things, perhaps by providing food and shelter for animals, and shade which excludes other plants.

In attempting to solve these problems we again use the experimental approach and invent an explanation or **hypothesis** to provide a likely answer to our question. We then test it by controlled experiment. Two situations are compared keeping the influences on each identical except for the one factor being studied. We do this to eliminate all other factors because animals and plants are affected in so many ways by their surroundings. The normal situation is the **control** and the other is the **experiment**. By comparing the two we can see if the experimental factor makes any difference.

Sometimes we find natural experiments in the field and only need to recognize them. Fig. 12.6 (p. 180) provides an example in which the sweeping of seaweed over smaller organisms is investigated by comparing the 'swept zone' with the unswept area outside it.

One observation which led to hypothesis and experiment was that the cabbage aphid, *Brevicoryne*, was found only on certain plants. As these belonged to the family Cruciferae whose members often contain mustard-oil compounds, the suggestion was made that it was these substances which induced the insect to remain on the plant. This was tested by placing aphids on two groups of broad bean plants – for the bean is not normally attacked. In the experimental group the leaves were fed with sinigrin, which is a mustard-oil glucoside. The control group lacked sinigrin. It was found that the aphids would only stay on the leaves which contained sinigrin, so the hypothesis was shown to be correct.

In the above example a field problem was solved by a laboratory experiment, but it is sometimes necessary and convenient to make the experiment in the field. A good instance is provided by Farrow's studies on the effects of rabbit grazing. Study of grass-heath in Breckland (Suffolk) revealed decayed ling roots in the underlying soil, suggesting its recent presence, though the nearest ling com-

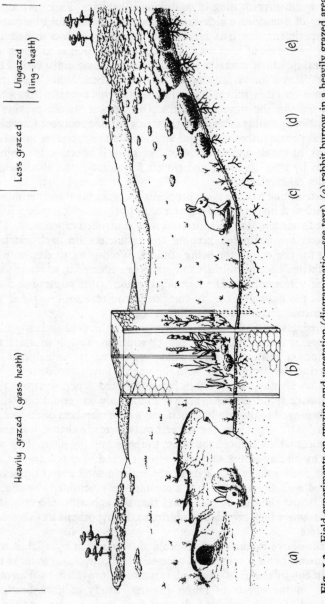

**Fig. 1.2** Field experiments on grazing and vegetation (diagrammatic—see text). (*a*) rabbit burrow in a heavily grazed area (grass-heath); (*b*) wire enclosure extending underground; (*c*) dead ling roots under the grass-heath; (*d*) nibbled and stunted ling; (*e*) healthy full grown ling. Extreme top left and right; pine woods are gradually replacing the ling

Heavily grazed (grass heath)

Less grazed

Ungrazed (ling - heath)

munity was some distance away. An examination of the vegetation between the two areas showed a progressive change which appeared to be associated with rabbit grazing. Farrow tested this by setting up rabbit proof enclosures in the transition zone and keeping a close record of changes in the vegetation. The control for this experiment was provided by the vegetation immediately surrounding the enclosure. Within the enclosure a succession of changes took place which was the reverse of changes observed in the control area. The grass-heath plants grew more vigorously but were soon replaced and dominated by ling which in turn was gradually replaced by pine or birch trees. This experiment thus not only confirmed Farrow's hypothesis, but provided further information showing how woodland entered the succession:

$$\textit{Birch-Pine woods} \xrightarrow[\text{ungrazed}]{\text{grazed}} \textit{ling-heath} \xrightarrow[\text{ungrazed}]{\text{grazed}} \textit{grass-heath}$$

Every biologist needs to know how to carry out original investigations like these because biology is the study of life in action. A set experiment copied in all its details from the textbook is not sufficient, because the essential elements of planning and discovery are missing. Field studies are particularly valuable because every habitat provides problems for which there are no textbook answers and which can only be solved by accurate observations and carefully planned experiments. To be successful at this, the biologist must cultivate a flexible and open mind because it is rarely possible for him to plan his research in detail for more than a few steps ahead. Programmes constantly need to be adjusted and ideas must be revised as new information becomes available.

Progress in field biology is more rapid when there is free exchange of ideas and information. Group projects and field society gatherings provide excellent opportunities for discussing one another's work, particularly when this is presented in writing or as talks. Ideas sparked off by questions and comments often form the basis of new discoveries; secrecy hinders progress.

# 2 The Elements of Ecology

Fieldwork is above all concerned with the relationships of organisms to their natural environments; the study of these is known as **ecology**. As an example of what this involves, we may consider the study of a common aphid (greenfly) feeding on its host, a tree. Our first interest will be to record where and in what numbers and forms the aphid occurs. After this preliminary survey, an attempt might be made to keep colonies in the laboratory and watch their behaviour under different conditions. Observations in the field should be repeated at intervals to include not only changes during the life history and fluctuations in the population, but also changes taking place in the surroundings. Typical questions to answer are:

When do aphids first appear on the leaves?

Where do they come from, and how do they find their feeding sites?

What are the causes and incidence of death throughout the year?

Where and in what form do aphids spend the winter?

Gradually a picture of the life-history, behaviour and relationships of the aphid will begin to emerge, and it will be based not on one individual but on many. In particular, the role the aphid plays in the general community of animals and plants becomes clearer as the investigations begin to include other organisms.

From this short example it will be obvious that the relationship between an organism and its environment is much more complex than might at first appear. Thus the organism must be considered not only as it is when first recorded but also as it was during the other parts of its life history, possibly in quite different places. Thus to understand why a plant or animal is found in its present surroundings we need to know how, when, and where it started life, what its needs were at that time, and what competition it has had to overcome in order to survive. In talking about environment, we are not concerned with the general surroundings just as *we* see them, but with the particular aspects that are important to the organism and therefore constitute its **effective environment**. The importance of these aspects changes with external factors – a tree may mean shelter in heavy rain but have little significance at other times. The internal state of the individual is also important – a pile of seeds means more to a starving sparrow than to a well fed one, and potential nesting sites are more important to birds in the breeding season than at other times. Two creatures found in the same kind of place usually have different effective environments because of their different needs. Thus voles and shrews in the same rough grassland are in different effective environments because voles feed on grass whilst shrews eat insects and other small invertebrates.

The field biologist soon learns that animals and plants must also be studied as populations because the group shows properties which cannot be expressed by isolated organisms. Courtship and parental behaviour are simple examples of activities which may involve single pairs or much larger groups like gannet colonies. Such activities by isolated individuals are seen only rarely, as when a zoo penguin tries to incubate a bun, or a peacock displays at a human being. On the other hand, populations may show group behaviour and evolutionary patterns very similar to those associated with individuals. Many other processes are based on the group and their variety, complexity and importance are a great challenge, because they bring together such different subjects as physiology, evolution, agriculture and sociology.

The most obvious group is the **species**, usually kept together by interbreeding with its exchange of hereditary factors or genes. As a consequence of this the species is said to have a common **gene pool**. However, the species does not have the uniform distribution that might be expected, but is divided into many subsidiary **populations** separated from one another so that they rarely interbreed. Each of these small breeding populations is called a **deme**. The isolation of some rabbit warrens ensured their survival when most of the rabbit population was wiped out by myxomatosis; other examples of demes include such groups as the trees in a beechwood, sticklebacks in a pond or a house-martin colony, all of which rarely interbreed with similar groups.

Demes of different species overlap and interact to form the distinctive pattern we recognize as a **community**. This is a well-defined assemblage of plants and animals, and clearly distinguishable from other such assemblages. Typical examples are a wood, a sand-dune or a pond. Communities are often named after the dominant plants, as in oak-wood and grass-heath.

Green plants are the **primary producers** of the community, transforming the energy of sunlight into chemical energy by building up complex organic compounds from simple ones. All animals depend ultimately on this chemical energy stored by plants, either directly, or indirectly by preying on herbivores. Thus aphids which feed on plant tissues provide food for ladybird beetles, which in turn may be eaten by birds or mammals. In such a **food chain** there is wastage at every link because some food is broken down to provide energy for vital processes and movement. As a result there is not nearly enough food for the predator at the end of the chain to keep its numbers as high as those of the herbivore which is a **primary consumer**. The effect is exaggerated by the fact that most predators owe part of their success to being larger than their prey. Thus there is often a gradation in size running in the opposite direction to the gradation in numbers. Aphids are smaller and greatly outnumber the ladybirds which feed on them; ladybirds are smaller and more numerous than their own enemies. The relationships of a food chain can be represented by a **pyramid of numbers** or a **pyramid of mass**. The steps in the pyramids are known as **trophic levels**.

The pyramid shown in Fig. 2.1 is much simpler than the true situation in nature because aphids form part of the diet of many different organisms and so enter quite different food chains. In the same way one predator hunts a variety of prey and may itself fall victim to parasites or predators. Different food chains are therefore

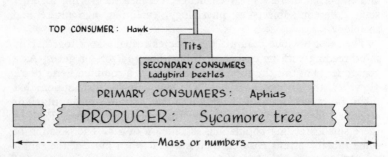

**Fig. 2.1** Pyramid of mass. When represented by the total weight or *biomass* of organisms at the different levels, size effects are eliminated. Aphids are much smaller and more numerous than sparrow hawks, so that a pyramid based on numbers alone would have a much broader base than the one shown here

linked through species which they have in common. These relationships are represented graphically as a **food web** in Fig. 2.2.

Each animal species has its own typical food relationships with other species, so in a given community at a given time each is said to have its characteristic **food niche**. Caterpillars and aphids may feed on the same plant leaves but their food niches differ because one is a chewing insect and the other a sap-sucker. As the seasons advance, changes in plants and animals inevitably result in altered feeding habits so that the food niches will also change. The different food relationships give rise to structural and behavioural adaptations for hunting, feeding and escaping. These include specialized mouthparts, armour and behaviour patterns which aid concealment.

Although the food web indicates routes for **energy flow** through the community, it does not show how *much* energy flows, or is lost or gained

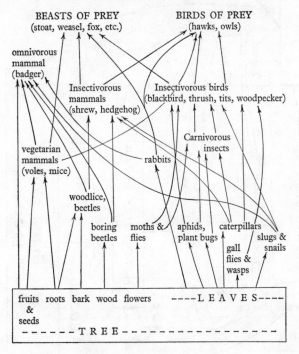

**Fig. 2.2**  Simplified food web

at any point. A deeper insight is always gained by *measuring* energy relationships (see Fig. 2.3). There is great wastage; only about 10% of the energy available to one trophic level passes on for possible use at the next. The proportion of energy passed on in this way is a measure of **ecological efficiency** and may vary from level to level. Wastage at herbivore level may be due to only part of the available plant material being eaten, with only some of that being fully digested and assimilated; even then much of the assimilated food is used to provide energy for vital activities. The loss of about 90% of the energy at each stage makes it easy to understand why food chains are usually short and those with six links or more are rare. (See also p. 232.)

The biomass of a population at one moment in time (the **standing crop**) does not always give a good indication of its significance in energy flow. What matters is the *rate* of production, or **productivity**, of new edible material at that trophic level. The standing crop of plant plankton in the sea is small compared with that of a land-plant community. However, plankton reproduce so rapidly that productivity is much higher, and total energy assimilated into the community over and above

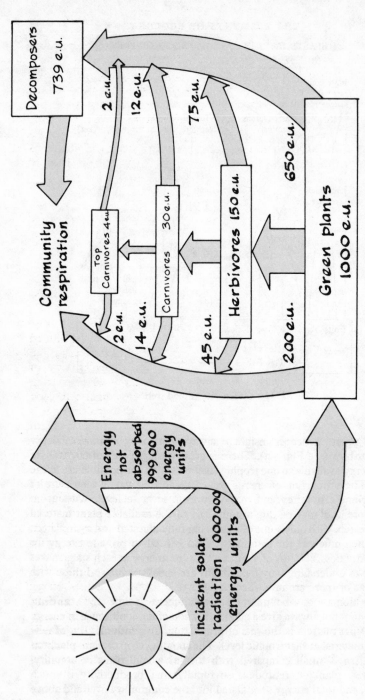

**Fig. 2.3** Energy flow in the community. Much of the solar energy reaching the plant is lost by reflection and photosynthetic inefficiency. Ascending arrows on the left represent energy losses in respiration; corresponding arrows on the right represent losses, through death, to decomposers. The sizes of boxes, thickness of arrows, and figures of 'energy units' (e.u.), suggest the general pattern of energy flow. In real life, the difficulty of estimation will mean that figures do not add up so neatly

that used in respiration during a set period, the **net primary production**, may be equal or even greater.

Energy flow is not the only aspect of community organization, for organisms may modify the non-living environment and provide or deny shelter to one another. A bee colony in winter clusters together but continues sufficient activity to keep its temperature above that outside and thus ensures group survival where the individual would perish.

Living organisms often depend for success on their ability to respond to unfavourable conditions in such a way that harmful

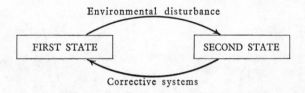

**Fig. 2.4**  Diagram showing how the feedback mechanism of a homeostatic system maintains the first (normal) state in face of disturbance

effects are checked or even reversed. Maintenance of a limited range of conditions in this way is known as **homeostasis**, and homeostatic mechanisms may be found at almost every level of organization from individual cells upwards. The levels which concern us most as ecologists are organism, deme and community. A good example of a homeostatic mechanism operating at organism level is seen in the way mammals are able to control their body temperatures and maintain them between quite narrow limits. When mammals are too cold they begin to shiver, and the muscular action of shivering generates heat which checks the fall in body temperature. If they are too hot, they become less active and sweat or pant to lose heat by evaporation. This is so effective that quite small departures from these limits in humans provide clear indications of disease. Many animals which lack these particular mechanisms still manage to keep their body temperatures between safe limits simply by moving to warmer or cooler places. This illustrates the fact that homeostasis of body processes is often achieved through **control behaviour**. Thus the study of behaviour (ethology) helps the animal ecologist because animals often reveal through their behaviour those parts of their effective environment which are important at any given time (Fig. 2.5). Control behaviour also provides information about the animal's immediate needs for maintaining both internal and external conditions at a satisfactory level. Such requirements include food, water and shelter as well as warmth and coolness.

The non-living or **physical environment** of air, water and soil

minerals is immensely important. The physical environment and
living community function together as an ecological system, or
**ecosystem**. The way in which such a system operates is well illustrated by the cycles of essential elements like nitrogen, carbon and
phosphorus which pass to and fro between the living and non-living
parts of the ecosystem. All these elements are taken from the soil or
atmosphere by the growing plant to form essential compounds.
Through food chains these may become incorporated in animals, but
eventually death and decay return them to the non-living environment
in the soil and air.

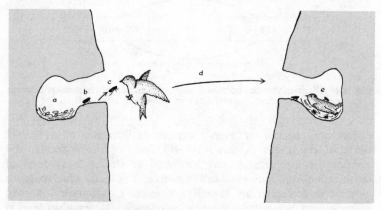

**Fig. 2.5** Homeostatic behaviour correcting food needs in sand-martin fleas.
(*a*) Flea cocoons in deserted nest. (*b*) Starving fleas move towards the light at
the entrance of the nest burrow. (*c*) Fleas sitting at the entrance jump when
passing birds throw a shadow on them. (*d*) The bird carries them to its nest,
where (*e*) they feed on the blood of nestlings and adult, in response to olfactory
and temperature stimuli

Just as each species has a food niche, it is also said to have a
**habitat niche**. An organism at any specific moment has a particular
effective environment and its habitat niche can be thought of as the
sum total of its many effective environments throughout life. We may
speak of the habitat niche when referring to an individual or the whole
species. Though a plant remains in the same place for the greater
part of its life, the effective environment of a seedling is quite different
from that of a full-grown plant. The effective environment changes
again during the reproductive phase with its dependence first on
pollination agencies and then on fruit and seed dispersal mechanisms.
Its habitat niche is the sum of all these effective environments and is
based on one place. In contrast, an animal as it grows up often *moves*
to a new part of its habitat niche. This is shown particularly well by

an insect like the mosquito, whose larval stages live in water and feed on small particles whilst the adult mosquito flies in the air and sucks blood. In a similar way eggs, nestlings, and adult birds have different relationships with their surroundings, and each of these relationships forms part of the habitat niche for the species. The term 'niche' on its own is much misused in ecological writing to mean food-niche, habitat-niche, habitat or microhabitat. This has caused much confusion so it is essential to define terms exactly. Habitat-niches are not specific places whereas habitats and microhabitats are. Field biologists use the term **habitat** to denote any specific area which is inhabited by plants and animals and therefore worth studying; within any particular community there are many such habitats. Thus habitats and microhabitats provide effective environments which form parts of the habitat niches of many different organisms.

In any one area there are only a limited number of places where the habitat niche requirements of a particular species are met. As time passes, changes render some of these places unsuitable whilst new and more acceptable ones are formed elsewhere. Cattle dung deteriorates with time, so animals living in cow-pats need stages in their life histories in which they can move to new droppings. In this way it is characteristic of all animals and plants to show some degree of **dispersal** – movement away from a population centre. Such movement often commences long before niche conditions become unsuitable, with the result that suitable new areas are colonized quite rapidly. Dispersal sometimes continues over a long period, but in many life histories there are sudden bursts of dispersive movement associated with structural adaptations such as winged generations in aphids, or seed and fruit dispersal mechanisms in plants.

Dispersal must not be confused with **dispersion** which describes the location of individuals within a population relative to one another and to features in the environment. A species tends to be dispersed in clumps rather than uniformly and the pattern alters with changes in the general environment. Dispersion will also depend largely on the composition of the population – the numbers in each age group and the gains and losses which are taking place, by immigration and reproduction in the first case, and emigration and death in the second. The study of changing populations is known as **population dynamics**, and the fluctuations associated with the changing seasons, weather and food supply provide particularly fruitful lines of investigation. Some changes form part of a regular annual pattern, whilst others are obviously related to population density – overcrowding may result in disrupted or migratory behaviour patterns and less food for individuals but more opportunities for predators or parasites.

Changing dispersions within a community often bring the same or different species into **competition** for limited resources of food, light

or shelter. Plants which cast heavy shade are said to compete successfully for light since few other plants can establish themselves beneath
them. Where two animal species search for a limited supply of the
same food, the population of the less efficient animal begins to dwindle
through starvation, migration or reduced birth rate. Within a species
**aggression** may result in more efficient use of habitat resources.
Threatening display and song by birds such as the robin help to establish territories which provide adequately for survival and breeding.
Notice that it is not enough to say that two individuals or species are
competing; it is essential to say what organisms are competing for, and
how they set about it.

Partly as a result of competition for light, the plants of a community
are usually stratified in fairly clearly defined layers (Fig. 2.6). Thus in
an oakwood the canopy of the **tree layer** is provided by oaks, the
**shrub layer** may be dominated by coppiced hazels and below these
there is a **field layer** of herbs. On the soil there may be a **ground
layer** dominated by mosses. Stratification may still be important
where all the plants are herbs; for instance tor-grass dominates rough
grassland and provides sheltered conditions for smaller plants. Many
of the **habitat factors** described in Chapter 6 influence the stratification of communities and you can discover some of these by field
observation and careful thinking.

We have seen that communities are not static but are continually
changing. In a stable community the changes are mainly cyclic and
are associated with seasonal variations in rainfall, temperature and
day-length. To help describe these changes, we distinguish five
**Seasonal Aspects** in the annual cycle. **Winter aspect** occurs when
most plants are dormant – trees and shrubs are leafless and herbs are
represented only by seeds or perennating structures. **Prevernal
aspect** follows when the first flowers (primrose, celandine, wood
anemone) appear but trees and shrubs remain leafless. During April
the **vernal aspect** commences with the bursting of buds so that trees
and shrubs are in young leaf and cast medium shade. Herbs likely to
be in flower at this time include bluebell, dog's mercury and greater
stitchwort. During the **summer aspect** the trees and shrubs are in
full leaf, casting a deep shade so that few herbs are in flower. The
cycle ends with the **autumnal aspect** when the trees and shrubs
lose their leaves, herbs die back and many fungi are in evidence.
These terms are most useful in relation to woodland which is discussed in Chapter 8. Seasonal fluctuations in animal populations due
to death, reproduction and migration are associated with the changes
in vegetation.

Distinct from these seasonal changes, some communities alter over
the years so that successive communities displace one another in
turn. The developmental history of a plant community is called **plant**

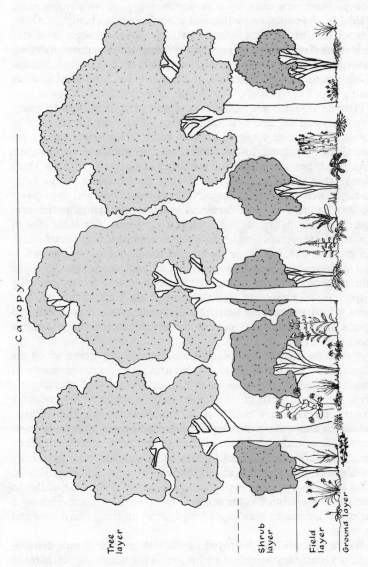

**Fig. 2.6** Stratification into four layers in a wood

Canopy

Tree layer

Shrub layer

Field layer

Ground layer

**succession,** and is further defined as **primary succession** when it starts with the colonization of a bare new soil by a **pioneer population.** Most new soils arise by accumulation of windblown sand, of tidal or river silts, or by removal of surface soil in landslips. Other new soils are products of industrial activity like the dumps associated with china clay and coal mining or the deep subsoil exposed by opencast mining. All are characterized by the absence of organic remains which would provide nutrients, soil stability and a favourable water status.

Plant successions are named according to the conditions in which they start, and fall into two main groups. Where conditions are very dry as on bare rock or in windblown sands, the pioneer species are specialized to withstand water shortage and are known as **xerophytes.** Such plants provide greater soil stability and a degree of shelter which allows other plants to enter and change the character of the community. Each community further affects conditions and thus paves the way for its own replacement. A succession of communities produced like this on dry soil is known as a **xerosere.** The other kind of primary succession begins in shallow water with pioneer plants which can grow in waterlogged or submerged conditions and are known as **hydrophytes.** Once again a succession of communities results as the habitat becomes drier when accumulated plant debris gradually raises the soil above the water surface. This kind of succession is called a **hydrosere.** Any particular plant succession may be referred to as a **sere** and each identifiable community in the succession is a **seral stage.**

Both xeroseres and hydroseres lead eventually to stages which are intermediate or **mesophytic** in character. The common plants of meadows, hedges and woodlands are all examples of mesophytes. From this point both kinds of sere develop similarly to produce a **climax community** which is in equilibrium with the climatic conditions and will not develop further unless these change. In most parts of Britain the **climatic climax** is deciduous woodland. However, continuous interference by man or his animals more frequently establishes the equilibrium in a different way termed a **biotic climax,** as for instance when intensive grazing suppresses the growth of shrubs and trees. No climax community is completely uniform because a variety of influences may cause irregularities in the pattern here and there.

Where vegetation is destroyed by burning or other drastic disturbance, a **secondary succession** follows. This begins on soil already present as distinct from the bare rock or mud at the beginning of a primary succession. In secondary succession the starting point may be felled woodland, burned heath or railway embankment, fallow ploughland or even a garden bed. The early stages are very different from those of a primary succession, and the course of succession is shorter. In general, however, the colonizing plants again undergo successive replacement and eventually re-establish the climax vegetation. **Subseres** of this kind are common and provide good opportunities for studying succession. On a much smaller scale, the changes taking place in microhabitats (cattle droppings, a rotting log, a temporary puddle) constitute a **microsere.**

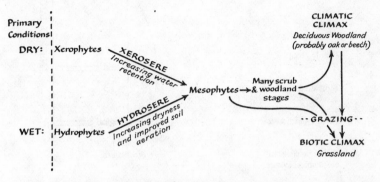

**Fig. 2.7** Diagram of succession

Plant succession may be mirrored by **zonation** on the ground. Thus on sand-dunes it may be possible to see all stages of a primary xerosere – from the newly-colonized fore-dunes to the biotic climax grassland usually present behind the dunes. The successive stages of hydroseres may similarly be seen in marshland at the edge of a lake, or on salt marshes in estuaries. Where zonation is seen, care must be taken over its interpretation; zonation on the seashore is related to tidal exposure and is quite unconnected with succession.

By now it will be obvious that ecology is concerned with a great range of processes operating at many different levels. In field biology we are mainly concerned with higher levels of organization such as communities, demes and individuals. Not only do all these depend on the kinds of interaction we have been considering but also upon the lower levels of organization such as cells, tissues and organs. These determine the potentialities of individuals and groups for responding and adapting themselves to new situations. Thus a plant's ability to grow in dry conditions is largely determined by the nature of its stem and leaf anatomy, and an animal's survival in similar conditions may depend on being able to conserve water through having a waterproof shell or cuticle, enclosed respiratory organs and insoluble excretory products. It follows that an understanding of many of the structures and processes inside organisms is essential if we are to appreciate

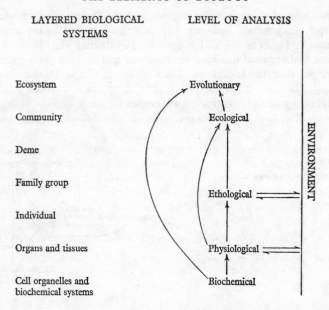

**Fig. 2.8** Hierarchical or layered arrangement of biological processes. Arrows show interactions between levels and with environment

why they live where they do, and why they behave in the way they do.

The ecologist should never be afraid to think broadly and along his own lines since, even if it involves a few false starts, this is how the frontiers of knowledge are eventually advanced. However, new ideas only come after close study of the actual material and the beginner must first learn to collect and identify his plants and animals.

# 3 Collecting and Identifying

Reliable collecting methods and accurate identification are essential preliminaries to any ecological survey and for various reasons these are usually easier for plants than for animals. Most plants are photosynthetic, and their need to receive light means that they are positioned where they can be seen. In contrast, animals are frequently camouflaged or hidden in dark crevices. Apart from being more conspicuous, land plants in Britain include fewer species than animals, so that books used in their identification are readily available and comprehensive.

## Collecting: how and where to look

It is usually best to focus attention on a limited area which can be thoroughly and systematically searched. The outlining of small areas in grassland by means of wire quadrat frames may reveal inconspicuous plants and insects which would be missed in a general and superficial survey. Such frames can be used in a variety of situations such as the bark of trees or parts of the sea shore. Naturally defined areas like a rock pool, if not too large, can be searched in a similar manner. Again, one can specialize in searching microhabitats associated with particular organisms. Examples include *Laminaria* holdfasts, the inflorescences of umbellifers, composites and other plants and birds' nests.

When searching for animals no possibility should be overlooked. Swellings in leaf, stem or roots of plants often mark the presence of an insect or other small animal which is disturbing the normal growth processes of the plant. Such galls ('oak-apple', marble gall of oak, etc.) are frequently specific for the inhabiting organism, and may be brightly coloured. In both terrestrial and aquatic habitats stones should be lifted to search their under-surfaces and the ground beneath. Each such stone must be returned to its original position to avoid disturbing ecological relationships. This is particularly necessary on the sea shore but obviously less so on a scree. Old logs can be lifted in the same way, but in addition the space under the bark and all decaying parts should be searched. Fungi which cause decay can be identified readily only when the fruiting bodies appear in autumn.

## General collecting apparatus

Most of the following apparatus can be usefully taken to any kind of habitat; some of the more important uses are indicated.

HAND LENS – of medium power (×8 or ×10), useful for recognition and identification of small structures in both animals and plants.

BULB-PIPETTES – both wide and narrow mouthed, for transferring small aquatic organisms to suitable containers.

FORCEPS (tweezers) – for transferring medium sized, hard bodied organisms.

PAINT-BRUSH – to transfer delicate or soft bodied organisms, such as flatworms and aphids.

WIDGER – a very useful all-purpose tool, used for digging out plants, chipping off rock-living organisms, levering open crevices, excavating rotting wood, etc. A steel trowel may be used instead of a widger.

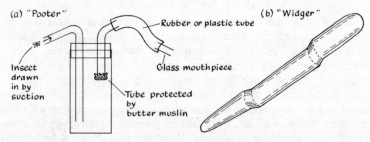

**Fig. 3.1**  *(a)* Pooter (aspirator). *(b)* Widger

POOTER (aspirator) – for collecting and transferring small insects and arachnids.

The pooter is constructed as in Fig. 3.1. The mouthpiece is placed at the end of a flexible tube so that insects may be captured at a comfortable viewing distance. The collecting-tube end of the suction tube is protected by butter muslin, so that the operator does not get an unwelcome mouthful.

SPECIMEN TUBES – some of these should be the same size (usually 75 mm × 25 mm) as the pooter tube, so that the empty tube can be quickly substituted for the full tube without having to transfer insects a second time. A second, smaller, size of tube (about 50 mm × 15 mm) is useful for small specimens which must be kept separate from others.

SIEVE – an all-metal kitchen sieve with handle is very suitable for sifting soil, mud, sand or pond water for small organisms. The lugs on the side opposite the handle should be bent close to the wire mesh so that they are out of the way.

SCREW-TOPPED JARS (Kilner jars, old pickle and peanut butter pots, etc.) – for collecting and transporting aquatic material.

GLASS BOTTOMED PILL BOXES – for collecting insects. The glass not only enables one to see the contents but makes collecting easier; the captured insect in its attempts to escape is attracted to the glass and away from the lid.

POLYTHENE BAGS (of varied sizes) – have practically replaced the metal vasculum for carrying flowering plants and are also useful for seaweeds or soil, where moisture retention is important.

CAMERA – for pictorial records.

FIELD NOTEBOOK – see pp. 32, 25, 41, Fig. 4.15.

In addition to these which may be carried by the individual, a shallow enamel dish or a white sheet (cloth or paper) may be taken by one member of a group, to provide a suitable background for the sorting of aquatic or terrestrial animals.

## Traps

Many shy or nocturnal animals are best hunted by trapping. Manufactured traps are readily available for catching small mammals, alive or

**Fig. 3.2** Longworth trap. (*Left*) placed in position, (*Right*) same trap camouflaged, mainly to avoid human interference

dead. The best live-trap is the Longworth small mammal trap which can be prebaited with the entrance locked open, so that small mammals can get accustomed to it before it is used for sampling. This aluminium trap is made in two parts – the *tunnel* with a trapdoor operated by a simple treadle, and the *nest-box* which clips firmly at an angle to the tunnel. When carrying, the tunnel fits neatly inside the nest-box so that

**Fig. 3.3**  Simple light trap; illuminated sheet with both vertical and horizontal surfaces

traps can be packed closely. The shiny aluminium makes the trap conspicuous so that it is necessary to disguise it if placed where people are likely to pass. This in turn requires a marking or recording system so that you can find the trap again – possibly a coloured marker set at constant distance and direction from the trap. It also helps if traps have been set at regular intervals on a line or grid (p. 69). The nest-box should contain plenty of dry hay for bedding and about a tablespoonful of whole oats – sufficient for a small rodent for a maximum of 24 hours; remember that the greatest dangers to small mammals are starvation and cold. Traps should be set with the nest-box tilted towards the tunnel (Fig. 3.2) so that moisture from condensation or urine will drain away. When sampling to estimate population size, at least 25 traps should be used; any smaller number will not yield significant results. 'Break-back' traps can be used where the animal is not required alive.

Many night flying insects and birds are attracted by lights and brilliantly lit surfaces. The simplest *light-trap* making use of this principle is an illuminated sheet arranged as shown in Fig. 3.3, with both vertical and horizontal illuminated surfaces. The most successful form of moth trap consists of a mercury-vapour lamp surrounded at its base by moulded egg-carton fragments amongst which the insects fall and settle. This type of lamp is far more efficient than a tungsten

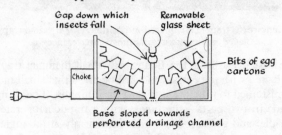

**Fig. 3.4**  Mercury-vapour light trap

lamp for collecting nocturnal insects, particularly moths, because the ultra-violet light it emits is particularly attractive to them. It has the disadvantage of being expensive and needing mains electricity.

Some aquatic insects can be attracted under water at night by an improvised light-trap using a hand torch inside a water-tight bottling jar as the light source, but this method is apt to prove expensive in batteries. The same trap can be used without a light; if left in a stream or pond for a day it may catch crayfish or small river fish which have ventured in and cannot get out.

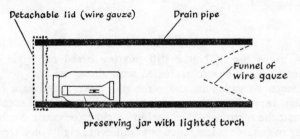

**Fig. 3.5**  Drainpipe trap

Migrating birds are often attracted by the lights of lighthouses and a number of bird observatories with large permanent traps are situated close to lighthouses on known migration routes. Information about these observatories and instruction in the art of trapping and handling birds should be sought from one of the recognized bird authorities (e.g. British Trust for Ornithology).

### Pitfall-traps

These are useful for collecting crawling or running insects. Such a trap is made by sinking a jam-jar so that its mouth is level with the beaten soil of a path or 'run' made by small mammals. The mouth

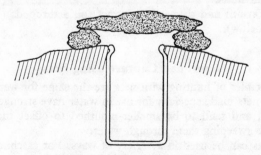

**Fig. 3.6**  Pitfall trap (see text)

should be protected by a roof of stone or wood. This can be arranged to exclude rain and larger animals like frogs or toads which might steal the insects. Pitfall traps should be arranged so that the ground falls away from them and there is no danger of flooding by drainage water. Decaying meat or animal excreta, placed as bait in the jar, will attract a variety of scavenging insects, whilst sweet and fermenting baits like over-ripe fruit, honey or jam will attract many of the insects that normally visit flowers. The traps should be visited frequently because the succession of insects attracted as decay proceeds may not survive together.

### Sticky-traps

These traps use as bait a stiff mixture based on crude sugar. Molasses or black treacle is boiled with about twice its weight of brown sugar to stiffen it, and some stale beer added as a further attraction. Immediately before using, a few drops of amyl acetate are added and the mixture is then spread on cloths or cards. Such traps are hung in suitable places (such as at different heights on a tree), and visited at intervals. As well as insects which actively seek the bait, the catch will include weak fliers such as aphids blown on to the trap by wind. It should not be forgotten that wounded or 'sappy' trees often provide natural sticky-traps which are worth frequent visits.

In contrast to attraction by baits, **repulsion methods** are often used to drive animals out of nests or burrows. Examples include:

(a) Extraction of earthworms by pouring a 0.5% formalin or 0.2% potassium permanganate solution over worm burrows. For quantitative work, the accuracy of this method must be assessed by digging and hand sorting on the same site.

(b) The use of smoke to drive out wasps, bees and other insects from their nests or from bark.

(c) Warm water to extract nematodes from soil (Baermann funnel, below) and to drive dor-beetles from their burrows or ants from their nests.

(d) Warmth and dryness to extract soil arthropods (Tullgren funnel, below).

## Pursuit and Capture using Nets

The principles of hunting with nets are the same for water as for air, though nets made specially for use in water have stronger frames and fabrics, and tend to be smaller-mouthed to offset the greater resistance to sweeping them through water.

Hand-nets can be used in a variety of ways. For catching insects in flight shorter handled nets can be wielded with greater precision.

A light, slightly flexible frame with a net of mosquito-netting or net curtain material is quite suitable. To avoid the escape of the larger and more powerful insects the bag should be at least twice as long as the diameter of the frame. The net must be used with a rapid sweeping movement, ending with a twist of the wrist to close the mouth.

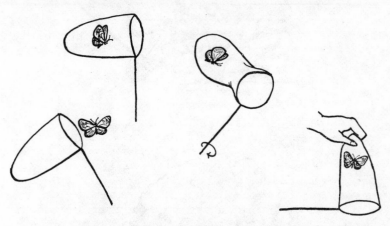

Fig. 3.7   Method of using hand net (see text)

If the net is now placed with its mouth on the ground and the tail of the bag raised, large insects like butterflies and moths will often fly up into the closed end making it easier to introduce them into a collecting box. Where the net is wide-mouthed and deep enough, the best way of capturing an insect may be to insert the entire upper part of one's body into the net; the use of glass-bottomed boxes for collecting has already been mentioned. Some insects are best captured by stalking them until they settle, then dropping the mouth of the net over them.

Another way of hunting insects with hand-nets is by sweeping. The net used here needs to be fairly robust, with a strong frame and a bag of fairly tough material such as calico. In sweeping the net is swept mouth first through grass or soft herbage, so that dislodged insects fall into the bag. The same robust net may be used under bushes or branches of trees beating upwards so that insects fall into the net.

A simpler way of 'beating around the bush' is to arrange a white sheet or newspaper under a branch, then beat the main part of the branch sharply with a stick. The active non-flying insects dislodged in this way are collected from the sheet using a pooter or are transferred direct to tube or box. An improved version is the Bignell **beating tray** in which the cloth sheet is stretched flat on a collapsible and portable lightweight frame. On a really flat sheet the insects are more

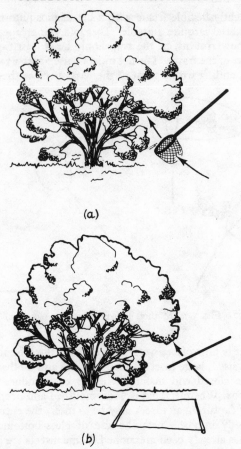

(a)

(b)

**Fig. 3.8**  Beating: (a) with net; (b) with stick

readily spotted and captured; many workers prefer dark coloured cloths to white, and this applies also to nets.

In water, moving animals can often be caught using smaller, moderately open-meshed nets with long handles. In small pools, the kitchen sieve mentioned earlier can be used to good effect in capturing small fish or shrimps. A sweep net for use in water needs to be extra tough because of the pressure on it when it is moved quickly. It is used in much the same way as on land, sweeping through water plants and sea-weeds, or beating upwards under floating weeds or rock ledges. The size of the mesh used is obviously important in determining the size of the catch. A shrimp net with four meshes to the inch will catch only organisms whose cross-section is more than one quarter of an inch across, but because of its open mesh it can be moved rapidly to capture agile creatures.

Plankton is collected using tow-nets made of bolting silk. In this material the threads are locked so that they do not separate under water pressure. The plankton net has about 40 meshes to the inch (even finer for plant plankton) and will only collect efficiently if moved at speeds as low as one or two miles per hour because a build up of pressure in the net creates turbulence at the mouth. This speed is slow enough to allow larger and more active creatures to escape

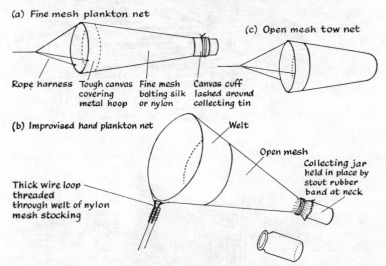

(a) Fine mesh plankton net

(c) Open mesh tow net

Rope harness  Tough canvas  Fine mesh  Canvas cuff
covering  bolting silk  lashed around
metal hoop  or nylon  collecting tin

(b) Improvised hand plankton net

Welt

Open mesh

Collecting jar
held in place by
stout rubber
band at neck

Thick wire loop
threaded
through welt of nylon
mesh stocking

**Fig. 3.9** (*a*) Fine mesh plankton net. (*b*) Improvised hand plankton net. (*c*) Open mesh tow net

without difficulty. For pond collecting, small plankton nets are made to fit ordinary hand-net frames. The tow-net used for collecting swimming organisms can be similar in overall dimensions to the plankton net, but having a much larger mesh is moved at greater speed.

By estimating the volume of water passed through the net, a rough quantitative determination of frequency per unit volume can be made.

The **drag-net** or **dredge** is a framed bag designed for hauling over a surface. For use in grassland or rough downland the frame needs to be constructed with runners, like a sledge, so that the net can be drawn along with the mouth of the bag facing forwards. This type of net is useful for quantitative sampling of caterpillars or other insect larvae which are readily dislodged from the vegetation. It can also be used as a static net in a stream to collect organisms deliberately dislodged from a position upstream; the frame makes it a useful alternative to the simpler screen of net pegged across a stream for the same purpose.

For a quantitative survey, a rectangular **quadrat frame** may be placed upstream and close to the mouth of the net. The stream bed

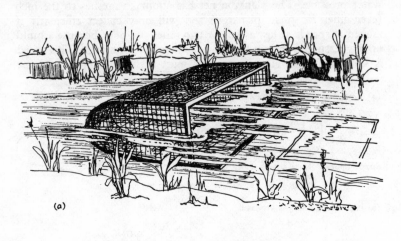

(a)

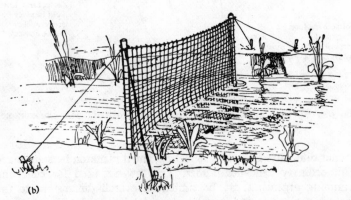

(b)

**Fig. 3.10**  Static nets used in streams. (*a*) Framed net and quadrat frame; (*b*) simple net screen

within this frame is then thoroughly disturbed to dislodge any creatures present so that they are carried by the current into the net.

A dredge for use under water can be of simpler construction but needs to be heavier, otherwise its forward motion and the buoyancy of water will combine to lift it off the bottom. In this case, the mouth of the net is held upright by the arrangement of the tow-rope harness. The lower edge of the net, which may come in contact with small rocks and other submerged obstructions, needs to be of specially robust construction and arranged so that it protects the material of the bag where it is attached to the frame. Alternatively, the net can be protected at the frame by rubber garden hose split along one edge or by a stout covering of sacking.

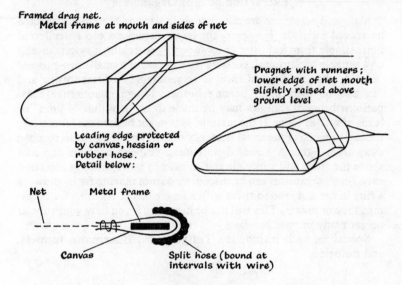

**Fig. 3.11**  Dredge nets

Useful additional items of equipment for use in fresh water are the grapnel, attached either to a stout cord or rod, a knife attachment and a 'spud' attachment, the last two for use on sectional rods. The grapnel and knife attachments can be used to obtain plant specimens which are

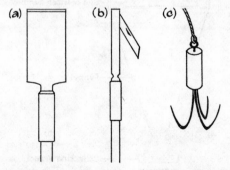

**Fig. 3.12**  Useful items for the collector: (*a*) spud; (*b*) collecting knife; (*c*) grapnel. (*a*) and (*b*) fit on to threaded handles

out of reach on trees and cliffs, but are most useful for obtaining inaccessible specimens of water plants. The 'spud' attachment has a sharp chisel edge which is useful for cutting off tough roots.

## Extraction of Soil Organisms

Many soil organisms are microscopic in size and need to be collected by special methods. However, the larger organisms can be collected quite simply from leaf litter or loose soil by searching it systematically on a suitable background, possibly with the aid of a small open-meshed sieve. A white rubberized sheet is very suitable for litter sorting and can also be used when beating bushes. Rough quantitative comparisons between samples may be made using 'handfuls of litter' or 'cans of soil' as units. If the material is brought back to the laboratory, use can be made of the usual tendency of litter and soil animals to move away from light. A hundred-watt lamp suspended about one foot above the sheet will stimulate the animals to move, making detection easy. Small organisms can be discouraged from moving far by drawing a ring in the soil around them with a finger before reaching for specimen tube or pooter. This will not be necessary if you have your tube or pooter ready in your hand.

Special methods include the **Tullgren** and **Baermann funnels**, and **flotation**.

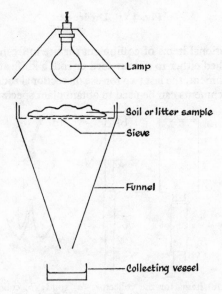

Lamp

Soil or litter sample

Sieve

Funnel

Collecting vessel

**Fig. 3.13**  Tullgren funnel

In the Tullgren funnel (Fig. 3.13) light and moderate heat are used to drive animals downwards in a soil or litter sample which is supported by a sieve. The animals fall through the sieve into a funnel which directs them into the collecting vessel. To be efficient the funnel must be perfectly smooth, and must not encourage condensation which traps the more delicate organisms before they can reach the bottom. For these reasons one of the best kinds of improvised funnels uses an elongated and therefore steep cone of smooth paper or card in preference to metal. The litter or soil can be placed in a kitchen sieve and warmed from above by an electric light bulb. The catch will be unable to escape if the collecting vessel contains liquid; water is most satisfactory but 70% alcohol or 4% formalin can be used for direct preservation. Damp blotting paper may be used instead of liquid to collect live specimens, but care must be taken to remove large predators like centipedes as soon as they appear.

The Baermann funnel is used for extracting nematodes and depends on the fact that they are heavier than water. The chief features of this apparatus are shown in Fig. 3.14. The soil sample is placed in the muslin bag and submerged in water. The nematodes leave the soil and sink to the bottom of the funnel from which they may be collected about 12 hours later, by drawing off the bottom half-inch of water. This kind of funnel is also used for extracting nematodes from plant material. Collecting efficiency can be increased by *gentle* overhead

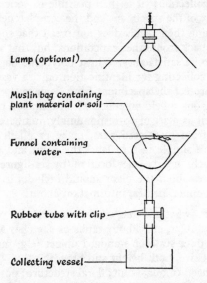

Lamp (optional)

Muslin bag containing plant material or soil

Funnel containing water

Rubber tube with clip

Collecting vessel

**Fig. 3.14** Baermann funnel

heating as in the Tullgren funnel (nematodes become heat paralysed and unable to swim at about 30°C).

In contrast to nematodes, many small arthropods can be separated from soil by flotation on a liquid like 25% salt solution, whose specific gravity is slightly higher than that of water. The soil is stirred into the solution and arthropods appearing at the surface are picked off by hand with a section lifter or paint brush. Flotation may also be used for extracting eggs, pupae and cocoons, not only of insects but of other invertebrates. This method is less useful for forest litter and peat which have large quantities of organic material that will also float. Insects can sometimes be separated from such floating material by adding an organic liquid like benzene which will 'wet' their waxy cuticles. The insects are recovered from the benzene whilst other organic material remains in the aqueous layer.

The methods just described are useful for qualitative investigations of soil and litter, but extraction is not sufficiently efficient for serious quantitative studies.

A comprehensive summary of methods for collecting soil animals is given in *Progress in Soil Zoology*, edited by P. W. Murphy (London, Butterworth, 1962).

## Collecting and Preserving Plant material

Special excursions should be made to gather plant specimens for the herbarium collection; it is then possible to concentrate entirely on the suitability of the specimens and the completeness of ecological records concerning them. This does not mean that specimens should not be brought back from other expeditions, but that such specimens are less likely to be satisfactory. Whilst polythene bags are perfectly adequate when collecting for identification only, a vasculum or large tin will better protect the specimens from damage.

Before setting out, a field notebook with numbered pages should be prepared, as well as a set of correspondingly numbered labels; price tags, available from most stationers, make excellent labels. In the field, it is then a simple task to label the plant and write complete notes about it including its exact location (by six-figure National Grid Reference). Notes about the plant should include, in addition to its scientific and common names, information about:

(a) HABIT: whether tree, shrub, annual or perennial herb, climber, scrambler, etc. Photographs or sketches are very useful; always include a scale or standard object (e.g. matchbox) as an indication of size, both height and breadth.

(b) FLOWERS: colour; scent; floral structure; position of flowering shoot; whether pendulous or erect; nature of inflorescence.

(c) FRUITS: degree of ripeness; colour and texture; dehiscent or indehiscent; smell; size and shape when fresh (it may shrink when preserved, or be squashed when pressed); any special method of dispersal.

(d) LEAF (and LEAFLET): colour; range of size and shape; arrangement (opposite, alternate, spiral, erect or horizontal).

(e) STEM: mode of branching; overall shape of plant; smooth, hairy or ridged; if woody, appearance and thickness of bark; storage and perennating structures.

(f) ROOT: taproot or fibrous root; storage.

(g) HABITAT: kind of community (e.g. woodland, grassland, sand-dune); associated species; frequency (see Chapter 4, p. 58); topography (Chapter 6); soil and water (Chapter 7); habitat factors (Chapter 6).

This probably seems a formidable list, but is not unreasonable if one bears in mind that *there is no need to waste time describing features still visible on the collected material*. The ideal is obviously to collect all parts of the plant including the root whenever possible; the few notes needed should be in simple unambiguous terms. Before collecting uncommon species make sure that plenty more are available – then take one specimen only. If the plant is rare, leave it and make a sketch or photographic record instead.

To preserve the plants they should be arranged carefully to show their distinctive features and pressed between sheets of absorbent paper. Cheap quality newspaper makes a good substitute for professional drying paper, and the specimens may be pressed under a pile of books or a weighted board. If the specimens are thick, they can either be split along their length and one half pressed, or they can be arranged with rolls of drying paper alongside so that the pressure is even. The success of pressing depends on changing the drying papers (each day if possible) until the specimens are dry. It is obviously going to save time if care is taken to use oven-dried newspaper or drying papers. If the specimen is placed in a thin tissue-paper cover at the start, the process of changing drying paper is made much easier.

When a specimen remains fairly rigid and feels brittle to the touch, it is sufficiently dry for mounting. The plant should be attached to a good white paper herbarium sheet by small strips of gummed paper (standard Kew size is 27 cm × 42 cm or $10\frac{1}{2}$ in × $16\frac{1}{2}$ in) or with a gum which will not penetrate and discolour the specimen. Cellophane self-adhesive tape is difficult to handle satisfactorily and should therefore be avoided. A botanical label must be prepared from the field notebook and attached to the bottom right-hand corner of the same sheet. This should contain information as in Fig. 3.15.

For completeness, structures such as fruits and seeds which were not available when the specimen was collected can be added in a similarly labelled packet attached to the bottom left-hand corner of the sheet.

Common name: *Jack by the Hedge / Garlic Mustard*
Scientific name: *Alliaria petiolata*
Family: *Cruciferae*
Locality: *Field path, Leckhampton, Cheltenham*
        *Map Reference: 945201*
Ecological notes: *West facing hedgerow;*
        *abundant and in flower*
Collector: *J. G. A. Plant*   Identified by: *J.B.T. Hedge*
Date: *4 May, 1965*   Specimen number:   *14*

**Fig. 3.15**  Herbarium label

Lower plants like Bryophytes and Lichens require different treatment – they are just left to dry, and are then housed in suitably labelled paper packets. If dried in a press, the pressure should be considerably less than is used for higher plants. If the packets or envelopes are of uniform size, with essential information written on the front, they can be kept like a card index. Mosses regain their flexibility and show their original characters when moistened, even some years after drying.

A herbarium collection can be arranged systematically, by Order and Family, or ecologically, by habitat. The second method is probably most rewarding for use in connection with field studies. If necessary, vegetational maps and transects can refer to a species by its herbarium sheet number until it is possible to get positive identification by an expert. Once identified, the herbarium specimen is always available for future reference.

Preservation of a representative collection of seaweeds is really worth while because there is no full identification key for marine algae suitable for use by amateurs and beginners. The tough and leathery Brown Algae are treated by first washing with fresh water to remove salt which, being hygroscopic, attracts moisture and adds to the problems of preservation. Very thick seaweeds can be softened by soaking for an hour in hot water, though the brown pigment fucoxanthin will wash out, leaving the alga green. For pressing, the alga is placed on a sheet of butter-muslin over several thicknesses of blotting paper or dry newspaper. Then a second sheet of muslin and more drying papers are placed above and weighted. Marine algae are usually mucilaginous, and the muslin stops them from sticking to the drying papers. Brown Algae generally need to be fastened to the mounting paper by means of gummed paper strips or by gum applied to the under-surface.

More delicate red, green or brown algae are dealt with as follows. The alga should be floated on fresh water in a shallow dish; this washes out the salt and makes it possible to see structure to the best advantage. A sheet of mounting paper is slid into the water and under the specimen, and then both are carefully lifted out on to a clean dish together. Here the arrangement of the specimen is completed using a camel-hair paintbrush charged with water. When suitably arranged, surplus water around the edges is removed with a cloth or blotting paper. Then a sheet of muslin is placed over the specimen and both it and its paper backing are placed between drying sheets and pressed. The natural mucilage on the seaweed is usually strong enough to fasten the specimen firmly to the mounting paper.

With all kinds of seaweed it is necessary to change the drying papers frequently, the first change being made after only a few hours.

Both higher plant and alga specimen sheets are best housed in paper folders. The best protection against mould is dryness so they should be stored in a dry place. Paradichlorobenzene crystals should be scattered in the collection from time to time to protect it from attack by insects or mites.

## Killing and Preserving Animal material

There is no credit to be gained from killing animals or taking birds' eggs merely to gratify personal 'magpie' instincts. Many animals are rare and some are legally protected so that to kill them is breaking the law as well as being morally wrong. The only real justification lies in the future use of specimens in furthering scientific knowledge; therefore decide how material is to be used and be selective when collecting. Never take rare animals, but record them by photographs or sketches instead.

As with plants, all relevant information should be recorded in a numbered notebook, and corresponding labels attached to the specimen. Where the latter is preserved in liquid, the paper label should be written in pencil and enclosed with it. When the specimen is finally placed in the collection, much of the notebook information can be transferred to a new additional label (on the *outside* of a jar containing liquid specimens). The information should include:

(a) SPECIFIC NAME – give common name also if in general use; give the name of identifying person.

(b) REFERENCE NUMBER – particularly important in liquid preserved specimens; the corresponding field notes should include details and sketches of colour in life (often altered during preservation).

(c) METHOD OF PRESERVATION (specimens in liquid) – information needed if part of the specimen is subjected to microscopic examination later.

(d) LOCALITY – sufficient description to find the place again; always give National Grid Reference. If collection is made during a survey by transect (see Chapter 4), indicate the 'station' at which the specimen was obtained.

(e) HABITAT – as for plants.

(f) SEX AND DEVELOPMENT – indicate condition: immature, larva, pupa, mature, pregnant, lactating, etc.

(g) SUPPLEMENTARY INFORMATION – this will depend on the group studied. In the case of mammals it might include dimensions such as lengths of: head and body, tail, hind foot, upper limb bones, skull, ear. If parasites are found, these should be mentioned with their own reference numbers. If the label is for a parasite it should provide name of host, position in host, locality where host was found.

Animals vary so much in size and structure that they cannot be preserved in a simple standard way like flowering plants, and the technique varies very much between different groups of animals.

Preparation of specimens is carried out in several stages: narcotization, killing, preservation and mounting.

## 1. *Narcotization*

This renders delicate animals less sensitive so that they will not be damaged or distorted by violent contraction when killed. This stage can be omitted with most arthropods.

The most useful narcotics are: (a) alcohol, added gradually to the surface of water containing the animals so that it reaches them slowly by diffusion and stupefies them. (b) Magnesium chloride; crystals of this are added to sea water containing marine specimens. Magnesium sulphate (Epsom salts) can be used in the same way. (c) Low temperature; cooling with ice or by placing in a refrigerator slows down the reactions of animals. This method can be used on its own or in conjunction with one of the other methods.

The time needed for narcotization depends on the size of the specimen, and may be as little as half a minute for small planktonic organisms.

## 2. *Killing*

This is often combined with the third stage, particularly if the killing agents are formalin or alcohol. Formalin (3–5%) used for killing and preserving should be buffered with hexamine (hexamethylene-tetramine, 200 g to a litre of 40% formalin) to prevent the production of free acid which will injure specimens, particularly those containing calcareous structures (echinoderms, arthropods, molluscs).

Alcohol causes shrinkage because it extracts water from specimens placed in it. Animal bodies contain large quantities of water so that an equal volume of alcohol will become diluted by one half. Animals must therefore first be placed in more than twice their own volume of 70–90% alcohol; a smaller volume of alcohol must be changed after 12–24 hours.

Terrestrial vertebrates can be killed using ether or chloroform. External parasites of birds or mammals can be collected readily from freshly killed specimens or from the killing jar. A large jar or air-tight drum is used as killing chamber, the entrance being covered with gauze over which is placed a pad of cotton-wool. The pad is then moistened with ether or chloroform and the vapour sinks through into the killing chamber, gradually becoming more concentrated and rendering the animals insensible. Care must be taken to ensure that the pad is *not soaked* until after the animals are unconscious as any anaesthetic dripping on the skin can cause unnecessary pain. For the same reason, the vapour concentration should be allowed to build up gradually, as animals suffer when placed directly into a high vapour concentration.

For insects it is usual to employ a killing bottle containing as killing agent potassium cyanide, crushed cherry laurel leaves or one of a variety of liquid chemicals (Fig. 3.16).

Of these methods, cyanide (1) is not recommended for complete beginners because it is a deadly poison and affects the colour of specimens; however, it is widely used by professional entomologists collecting certain insect groups. Cherry laurel leaf (2) is probably the best killing agent. Fresh dry young leaves are chopped finely, then placed in a cloth bag and bruised, causing them to give off the lethal gas (prussic acid). They are then placed in a thick layer at the bottom of a screw-top jar and covered with blotting-paper. Once prepared, this killing bottle will last up to eighteen months. The killing time is slightly longer than with other methods, but insects can be left in the bottle where they will remain perfectly relaxed for many hours, until it is convenient to set them. The third type of killing bottle (3) is constructed so that it can easily be recharged with liquid using a bulb pipette. Suitable liquids include chloroform, carbon tetrachloride, ethyl acetate and ammonium hydroxide. The first two have the disadvantage that they evaporate quickly, so that ethyl acetate is a more usual choice; it is a quick and effective killer, but mounting should be done soon after killing when the specimens are still in a relaxed and extended condition. Ammonia (full strength '880' which is unpleasant and dangerous to handle) is used by some collectors because it leaves butterflies and moths in a good condition for setting;

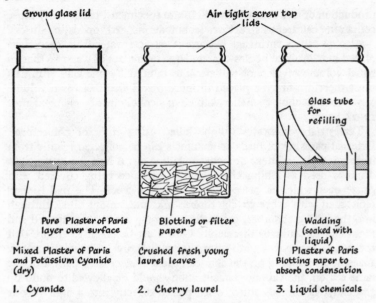

**Fig. 3.16** Insect killing jars, using (1) cyanide, (2) cherry laurel leaf, (3) ethyl acetate, or other similar liquid

however, it has the disadvantage of sometimes altering the colour of specimens.

Some larvae are more resistant to killing than adults, and may damage adult specimens if placed in a killing bottle with them. A good method for killing these larvae is to drop them into a small quantity of hot water (not boiling), then transfer them to alcohol after cooling.

### 3. Preservation

Animals killed with alcohol or formalin should be preserved in 60–70% alcohol or 3–5% formalin; alcohol is the better storage medium for general use. Variation in size involves the use of a wide assortment of containers. Wide-mouthed glass jars, with tight fitting glass lids on rubber washers (as used in fruit preserving), are best because it is possible to see the specimen and its condition, and to take action such as changing preservative when necessary. Metal lids, unless protected inside and out by a heavy coat of paint or by tinning, will soon begin to corrode. As already mentioned, a label – written clearly in soft lead pencil on good quality writing paper – should be immersed with the specimen so that it can be read through the glass of the container. (Labels on the outside are apt to come off or deteriorate.)

With birds and mammals it is usually necessary to preserve only the skin and skull. The skin is removed carefully, starting with a ventral incision. The terminal parts of limbs are retained, and possibly the entire limb skeleton as well; however, the muscles of the upper limb must be removed. The skin is preserved by rubbing in a powder containing equal proportions of borax and alum. The skins of birds and larger mammals are usually stuffed and arranged so that colour markings and other distinctive features are clearly displayed. Thus in birds the front part of the skull is left to show beak shape, and the whole skin stuffed with tow and sewn up ventrally with one wing and one foot exposed and sufficiently extended to show the main features. With mammals it is often possible to preserve head shape without incorporating the skull. A data label should be firmly attached to the right hind leg of the specimen; if the skull is preserved separately it should have a similar label bearing the same reference number.

Small mammals can often be skinned from one end entirely, either through the mouth or through a similar sized incision at the hind end. It is useful to keep a tin of the preserving powder at hand, and dust the

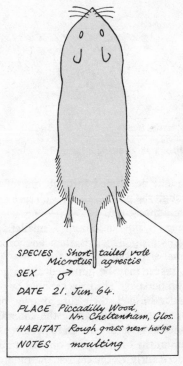

SPECIES  Short-tailed vole
         Microtus agrestis
SEX      ♂
DATE     21. Jun. 64.
PLACE    Piccadilly Wood,
         Nr. Cheltenham, Glos.
HABITAT  Rough grass near hedge
NOTES    moulting

**Fig. 3.17**  Small mammal skin mounted on card for preservation and filing

animal with it before and at frequent intervals during dissection because this will absorb animal fat which might otherwise discolour the coat. It is difficult to preserve the shapes of small mammals and it is often most useful to keep the skin stretched over a card which bears the normal data, the more important items such as species, locality and reference numbers being marked on both sides (Fig. 3.17). The card used for this must be rigid because the skin shrinks as it dries out. A detailed account of skin preparation is given in '*The Handbook of British Mammals*' (see p. 195).

Most insects have sufficiently rigid exoskeletons to allow dry preservation by pinning out and displaying.

The methods in most common use are:

1. *Direct pinning* with entomological pins through the thorax, spreading the legs and wings to display the maximum amount of detail. Care should be taken to ensure that the pin is inserted to one side of the mid-line. Large winged insects may be arranged with the wings of one or both sides extended on a setting board, the wings being held under pinned strips of paper until set.

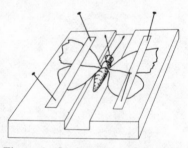

**Fig. 3.18**   Setting a butterfly's wings

Large beetles should be pinned through one of the elytra (wing covers) and not through the prothorax. In each case a data label should be attached to the mounting pin.

2. *Indirect pinning.* Using micropins, smaller insects may be fastened to polyporus strip which is then anchored to the display cabinet with a normal-sized pin.

3. *Carding.* The insect may be gummed on to good quality card (visiting card) or thin perspex. The insect should be placed on its back, gummed on its under-surface, then turned over on to the ready-gummed card and arranged by spreading the antennae and lining up the feet. The card should be large enough to protect as well as support. Data may be written on the card itself or on a separate data label. The mounting pin is placed through the card at the hind end of the speci-men. Carding on perspex makes it possible to examine under-surfaces.

As an alternative to the above, the insect may be fastened with gum to the apex of a paper triangle.

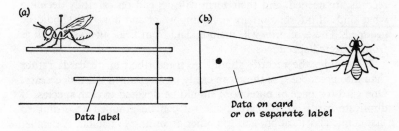

**Fig. 3.19** Mounting insects: (*a*) indirect pinning; (*b*) carding on paper triangle

As in the case of plants, specimens preserved in the dry need to be protected from mould, insects, mites and mice. This is effected by storage in completely dry conditions, use of an insecticide such as paradichlorobenzene, and provision of containers with tight-fitting lids.

## Organisms identified in food

Food preferences of many animals may be found by examining gut contents and by analysing faeces or the regurgitated pellets of predatory birds.

(*a*) *Gut contents.* Soon after death, these may contain recognisable remains of the last meal. Later, the more resistant remains such as hair, bone, or shell may still give clues.

(*b*) *Faeces.* These should be gently broken up in water, then washed through a fine sieve to recover such fragments as small mammal bones and hair, insect exoskeleton, snail shell, earthworm chaetae, and recognisable vegetable matter, all of which may be present in badger dung. Materials passing the sieve may be identifiable with a microscope.

(*c*) *Owl and hawk pellets.* These should be teased apart dry, and may contain the undigested bones, fur or feathers of prey. Look particularly for skull and lower jaw fragments; the grinding surfaces of teeth are particularly important in identification. Hair and feathers may be identifiable under the microscope.

References: SOUTHERN, H. N., Ed., (1963). *The Handbook of British Mammals.* Blackwell, pp. 141–147.

DAY, M. G. (1966) Identification of hair and feather remains in the gut and faeces of stoats and weasels, *J. Zool.* **148**, 201–217.

## Records

Detailed information from the field notebook becomes increasingly difficult to locate as data accumulates. More accessible permanent records are needed, and their form will depend on carefully deciding what kinds of information are most important and how they should be grouped. These records will include data from field surveys as well as from collected specimens.

If possible, the records should be transcribed on to cards rather than into notebooks, as these can easily be rearranged when necessary. One card (or page of notebook) should be devoted to each species. If duplicate cards are prepared, one set can be grouped according to taxonomic classification, and the other according to habitat. A detailed classification of habitats is given in *The Pattern of Animal Communities* by C. Elton.

When compiling records, it is obviously important to state how accurately species have been determined; a report based on unreliable identification has limited value. For most purposes it should suffice to give the names of identification books used, and a clear note about species which have raised doubts or difficulties.

| Species............................................................. | | | | |
|---|---|---|---|---|
| Date | Ref. No. | Locality | Habitat | Special Notes |
|  |  |  |  |  |

Fig. 3.20   Some suggested card index headings

## Identification

For the beginner, identification to species level is less important than identification to higher levels, and it is only with certain groups of flowering plants that identification to species level can be regarded as a reasonable possibility. In other groups, such as insects, it is more useful to start by identifying first to Order and later to Family, learning the characteristics of these groupings in the process.

The works recommended for use in identification are mainly wide in scope but limited in depth; many use illustrations which, though frowned on by professionals, do convey an impression which is often difficult to express in words. The field biologist *must* learn to use identification keys accurately, but the simpler works do enable him to make a start when he might otherwise be discouraged.

# 4 Survey and Recording Techniques

A clear picture of the organisms present and of their distribution in time and space is needed to reveal the ecological problems and show which of them can be investigated profitably. This involves identifying, listing, mapping, and counting organisms at intervals throughout the year.

Where possible, a preliminary survey should be made to record on sketch maps the limits of vegetational areas suitable for investigation by the methods outlined below. At the same time, it is essential to give careful thought to the form this investigation should take. What is the simplest way of classifying the major habitats? Which ecological events and processes are important and which can be ignored? How should they be recorded? Over what period must the survey extend? Remember that many ecological processes are slow. Will it be possible to maintain continuous records without interference – in other words, how suitable is the place chosen for investigation?

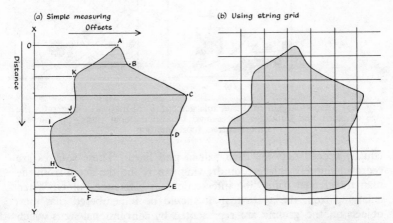

**Fig. 4.1** Direct mapping methods. (*a*) Simple measuring. Key points (A to K) are selected and marked. Offsets from the base line are measured and recorded, together with the distance along XY. (*b*) Using a string grid. Here the shape may be plotted directly on graph paper

## Mapping methods

### 1. *Simple measuring methods*

These involve setting out a measured base line with string and fixing important points by measuring 'offsets' at intervals. Using the fixed points as a framework, the rest of the map is filled in freehand. This method (or the alternative method using a grid of strings) is particularly useful for small ponds and rock pools and can be modified in various ways to suit circumstances (Fig. 4.1).

### 2. *Map enlargement*

Here the appropriate squares on a suitable Ordnance Survey map are enlarged on to drawing paper and detail is added in the field. The 6 in. map is the most useful, though adequate enlargements can be made from the 1:25 000 ($2\frac{1}{4}$ in to 1 mile) maps. Each 1 km grid square to be enlarged should be divided into 100 metre squares by drawing light pencil lines parallel to the grid lines (i.e. nine lines,

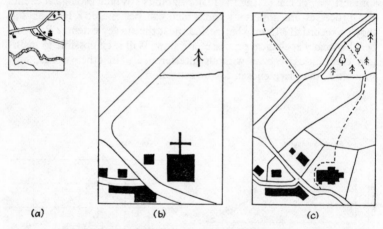

(a)                  (b)                  (c)

**Fig. 4.2** Map enlargement. Exact enlargement (*b*) of map (*a*) exaggerates the size of roads and buildings represented by conventional signs; (*c*) shows a more correct representation

equally spaced between each pair of grid lines). These squares are redrawn suitably enlarged on drawing paper, and the detail from the map transferred using line intersections as reference points. When enlarging from the $2\frac{1}{4}$ in map it should be remembered that most objects on the ground are represented by conventional signs which occupy more than their equivalent area on the map. Thus, when enlarging, roads, lanes, railways and paths must be sketched in much narrower than they would be by straightforward enlargement.

### 3. *Plane-Table survey*

The plane-table consists of a drawing board supported on a firm tripod and held by a central screw. Good 'ex-Government' tripods which can easily be adapted for this purpose are often available from photographic and scientific-instrument dealers. A simple spirit level and an alidade complete the equipment (Figs. 4.3 and 4.4). The alidade is a sighting ruler with a sighting slit at one end and a frame with a vertical wire in the centre at the other. This can also be improvised and should be made with its 'working edge' bevelled.

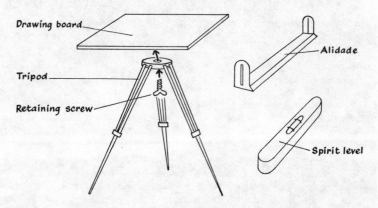

Drawing board

Alidade

Tripod

Retaining screw

Spirit level

**Fig. 4.3** Plane table and associated equipment

Unless the drawing board is very large, the drawing paper should be wrapped over the edges and pinned to the underside so that the alidade can be moved freely without being obstructed by pins.

The plane-table is used as follows:

**1.** A base line is set out on the ground and its length and compass bearing noted. The end points (A and B) are clearly marked (by pegs or flags on short sticks). Great care must be taken when selecting a base line.

**2.** Points which need to be fixed accurately on the map are selected and marked (e.g. with marker canes).

**3.** The base line is drawn to suitable scale on the drawing board, leaving ample space all round for completion of the map.

**4.** The plane-table is set up over point A and lined up so that the base line AB on the paper is pointing towards point B. The table is then levelled carefully, using the spirit level first along the board and then across it. When the table is level, the alidade is placed with its working edge on line AB. The centre screw holding the board is

**Fig. 4.4**  Mapping by plane table. *Top:* sighting with the alidade. *Bottom:* drawing board removed to fill in details of a rock pool. Note the flags and plastic clothes lines which have been used as markers

loosened slightly and the board accurately lined up by sighting through the alidade on to B. The screw is then tightened. Mark in the direction of Magnetic North (or record the bearing of B from A), using a prismatic compass if available.

**5.** Using a pencil point at A as a pivot against the working edge, the alidade is sighted on the selected points in turn, and faint lines (with identification marks) are drawn (Fig. 4.4, *top*).

**6.** The plane-table is then moved to point B, and the process repeated (level the board, align the drawn base line by sighting on A, check Magnetic North line with compass and draw lines in the directions of selected points using point B as pivot for alidade).

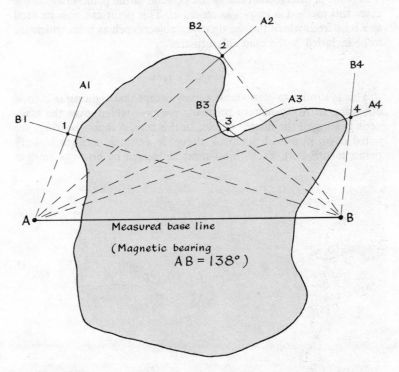

**Fig. 4.5** Mapping by plane table (see text). All lines are given identification marks as shown. Points 1 and 2 are located accurately (lines at right angles) and may be used to base the plane table for further sightings

**7.** The intersections of lines from A and B mark the positions of the selected points (Fig. 4.5).

**8.** Using these points as a framework, the map is filled in freehand (Fig. 4.4, *bottom*).

It should be noted that lines which intersect at right angles provide more accurate location of points than lines which intersect obliquely. This will involve moving the plane-table to one of the better fixed points to obtain lines giving greater accuracy at some intersections.

The plane-table can be used as an aid to map enlargement. With the enlargement on the drawing board, the table is aligned using a magnetic compass to set the grid lines, due allowance being made for the fact that grid north (on the map) is different from Magnetic North. Then the exact position of the plane-table is plotted by siting on two suitable points already marked on the enlargement, using the marks as pivots for the alidade and drawing lines back towards the observer. The point of intersection marks the position of the plane-table on the map; this method is known as resection. This point can now be used as a base from which to take sights on objects such as trees which are to be included in the map enlargement.

### 4. *Compass survey*

This is similar to plane-table survey except that sighting is carried out using an Army-pattern prismatic compass which gives the magnetic bearing of the distant object. In this case, compass bearings are noted down in a field notebook alongside a sketch map (to identify points mentioned). Each object must be located by bearings from at

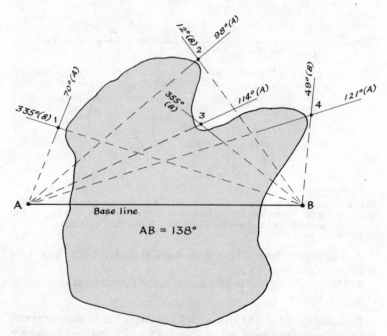

**Fig. 4.6** Mapping by compass survey. The same sights are taken as in Fig. 4.5, and compass bearings noted on a sketch in the field notebook. From this an accurate drawing is made

least two points, preferably so that the lines intersect at right angles. The fixed points can then be plotted on a drawing using a 360° protractor. All angles are measured clockwise; North is 360°, East is 90°, South is 180° and West 270°. If sighting is mainly from the ends of a measured base-line, the scale will be determined by the length of the base-line on the drawing board (Fig. 4.6).

The prismatic compass provides a useful and quick way of locating fixed points, though results tend to be less accurate than with a plane-table. It can be used to make rapid additions to maps made by enlargement or by plane-table (the observer's position is again located by resection ; in this case the bearings of two fixed points are converted to back-bearings by adding or subtracting 180°, after which the position can be plotted with a protractor).

By means of any of these methods a map of the area under investigation can be prepared. The amount of detail of flora and fauna shown will naturally depend on the scale, but in woodland for instance the map should show any important variations in the tree and other flora, paths (including paths made by animals such as badgers), some indication of contour and aspect and so on.

## Transects

Where there is some clear or suspected transition of vegetation and fauna, it is useful to make a detailed study along a line or *transect* which cuts across the area. Great care should be taken in choosing this line – it is better to start where many of the changes are evident at first sight. The line is marked out on the ground with string or, since it is more clearly visible, plastic coated clothes-line. If a transect is to be useful, the plant and animal studies must be accompanied by some investigation of habitat factors. The position of a transect (or any other detailed observation) should be marked accurately and clearly on the working map of the area.

The most common types of transect are dealt with in detail below, or illustrated in Figs. 10.1 (p. 145), 12.2 (p. 174), and 13.3 (p. 193).

### Profile transect

In its simplest form this consists of a record of plants occurring along the line together with a record of the changing level of the ground. For a short transect, changes in level can be recorded simply, using measuring tape and metre sticks. Plants are recorded every 10 cm, and the change in level recorded every 30 cm or at shorter intervals where there are sudden changes. The plant nearest to the point where a recording is to be made is noted together with its height (Figs. 4.7, 4.8).

Records in the field notebook should come under at least five headings:

TABLE 4.1

| Distance | Plant species | Plant height | Level | Notes |
|---|---|---|---|---|
| 0 | *Mercurialis perennis* | 25 cm | 0 | — |
| 10 cm | *Galium aparine* | 28 cm | — | Scrambling over *Mercurialis* |
| 20 cm | *Mercurialis perennis* | 22.5 cm | — | — |
| 30 cm | *Anthriscus sylvestris* | 38 cm | −23 cm | — |
| 40 cm | *Glechoma hederacea* | 12.5 cm | — | — |

and so on . . . . . . . . . . . .

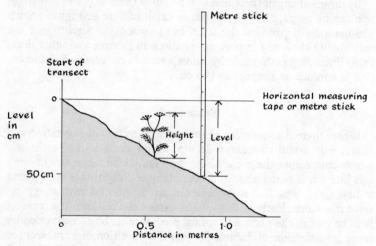

Fig. 4.7   Profile transect measurements

Note that scientific names are always used in preference to common names, as some common names are used for more than one species of plant. (There are over a dozen yellow-flowered species belonging to the genus *Ranunculus* – all of which might be loosely referred to as 'buttercup'). On the other hand, it is better to use a common name such as 'unidentified grass' or even code letters (Fig. 4.14) than to leave a blank if genus and species cannot readily be determined.

The selection of one plant at a time reduces the number of individuals recorded so that each plant can be conveniently represented by an arbitrary convention in a scale drawing. At the same time this introduces the possibility of error owing to a natural tendency to record the most conspicuous plant at each point. For this reason a profile transect should be accompanied at least by a *consolidated*

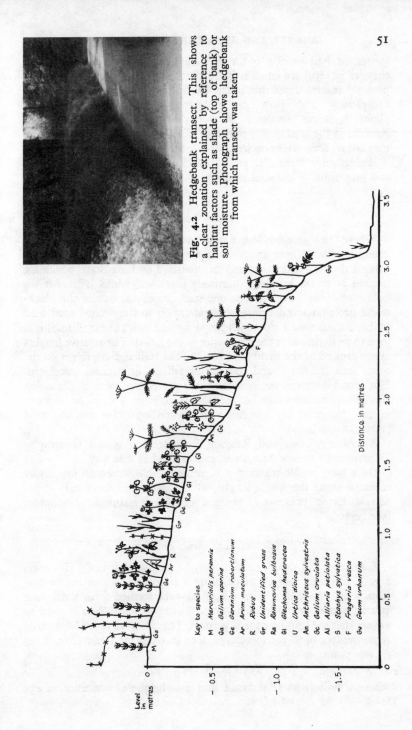

**Fig. 4.2** Hedgebank transect. This shows a clear zonation explained by reference to habitat factors such as shade (top of bank) or soil moisture. Photograph shows hedgebank from which transect was taken

Key to species

M   *Mercurialis perennis*
Ga  *Galium aparine*
Ge  *Geranium robertianum*
Ar  *Arum maculatum*
R   *Rubus*
Gr  Unidentified grass
Ra  *Ranunculus bulbosus*
Gl  *Glechoma hederacea*
U   *Urtica dioica*
An  *Anthriscus sylvestris*
Gc  *Galium cruciata*
Al  *Alliaria petiolata*
S   *Stachys sylvatica*
F   *Fragaria vesca*
Gu  *Geum urbanum*

Level in metres

Distance in metres

*species list* and usually by a *belt transect* also (see below). Where changes in level are small it is usual to exaggerate the vertical scale, possibly several times that of the horizontal. It should be noted that conspicuity varies between the vegetative and reproductive state; plants in flower are much more obvious than at any other time. Patterns in the community may also affect conspicuity – a plant which may attract little attention when evenly scattered becomes conspicuous if aggregated to form clumps or tussocks. The same is true of solitary and gregarious or colonial animals.

### Line of levels

Where the transect is long a more accurate method for determining changes of level over greater distances is needed. This is achieved using a simple Field Level and improvised Levelling Staffs which are marked at 20 cm intervals alternately black and white. The observer holds the Field Level against one staff (usually at one of the black-white junctions) and sights horizontally on to the second staff held by his assistant who moves his finger up and down the staff until it is seen to be in line with the cross-wire of the Level. The relative heights above ground of the Field Level on the one staff and the finger on the other staff are noted and the rise or fall in the ground calculated. Horizontal distances between staffs are determined with a measuring tape. The simplest method of recording is to note ' $+x$ cm' or ' $-y$ cm' as the change of level between one recording point and the next (Figs. 4.9 and 4.10).

An Abney Level and Ranging Rods from a school Geography Department's survey equipment can be used in this way.

On a long profile-transect it is impracticable to record organisms in detail along the whole length, so it is usual to record only at intervals, noting 'presence or absence' or making quantitative estimates (see below).

### Heights of tall shrubs and trees

High trees and shrubs cannot easily be measured directly. However, a good estimate may be made using a simple apparatus improvised from a 45° set-square, with a sighting-tube formed from a drinking straw fastened on with sticky tape. A simple plumb-bob is used to ensure that the leading edge is vertical. The observer walks backwards from the tree until the top can be seen through the straw (Fig. 4.11). At this point:

$$\text{Total height} = d + x,$$

where $d$ = horizontal distance and $x$ = height of observer to eye level.

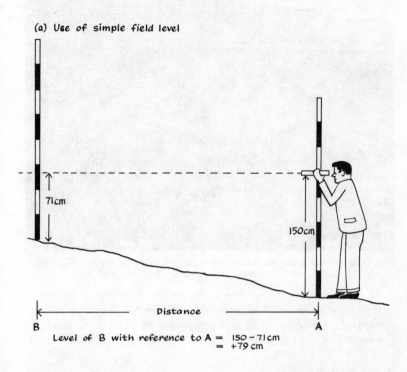

(a) Use of simple field level

71cm

150cm

Distance

B                                    A

Level of B with reference to A = 150 − 71 cm
= +79 cm

(b) Diagram showing construction of simple field level

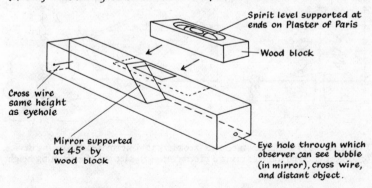

Spirit level supported at ends on Plaster of Paris

Wood block

Cross wire same height as eyehole

Mirror supported at 45° by wood block

Eye hole through which observer can see bubble (in mirror), cross wire, and distant object.

**Fig. 4.9**  (a) Method of determining change of level, using a field level. (b) Construction details of simple field level

**Fig. 4.10**   Using a simple field level to record a line of levels. (Woodbury Common. See also Fig. 10.1)

A slightly more useful instrument can be made using a protractor instead of a set-square, and this will enable an observer to measure the angle of a slope as well as height.

**Height measurement**

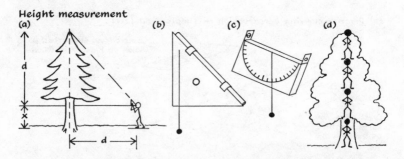

**Fig. 4.11**   (*a*) Height measurement (see text). (*b*) Improvised measuring device. (*c*) Improvised clinometer using a protractor. (*d*) Method of estimating height

If special instruments are not available, stand an object of known height (another person, a levelling staff, etc.) by the tree and estimate how many more objects of the same height would need to be superimposed to reach the top of the tree.

## Structural records of vegetation

The realistic diagram produced by making a short profile transect yields useful information about the structure of a community, and the relations of its members to one another and to habitat factors. The time-saving use of symbols proposed by P. Dansereau (1951, *Ecology*, **32**) may be just as useful for many purposes (Figs. 4.12 and 4.13).

All methods of recording have their limitations, and Dansereau has listed some of the advantages and disadvantages of this system. The most important points are given in Table 4.2.

TABLE 4.2

| Structural classification | System shows | System does not show |
|---|---|---|
| Life-form and size | Height of vegetation in each layer. Space occupied Woodiness (T, S) and herbaceousness (H) | Outlines of individual plants (e.g. shapes of tree crowns) Actual life forms Modes of branching and general habit |
| Function | Degree of deciduousness | Time of leaf fall Position of regenerating buds |
| Leaf shape and size | Six classes based on shape and size | Actual shapes of leaves Nature of margins, hairiness, position on plant (e.g. opposite or alternate) |
| Leaf texture | Four principal types | Relationship of texture to function (plants with sclerophyllous or succulent leaves are frequently but not always xerophytes) |
| Coverage | Coverage in terms of above classification Relative distribution of plants in each layer | Actual distribution of individuals as seen in conventional profile transect Plants of different layers according to same size scale |

## Line transect

A line transect (Fig. 4.14*a*) omits the heights and levels of a profile transect and simply consists of records of plants touching or covering the line all the way along or at regular intervals. The intervals are chosen to suit the vegetation, and may be as little as one inch in short turf. Total cover, the vertical projection of plant organs on to the line, normally exceeds 100% because stems and leaves often overlap. Though the length of interception of the **canopy** of a tree or shrub is recorded, it should be noted that canopy is not necessarily equivalent to cover, because useful light may still be reaching the ground.

**LIFE FORM**

| | | |
|---|---|---|
| T | ◯ | Trees (Tall woody plants, usually with single trunks) |
| S | ◯ | Shrubs (Shorter woody plants with several stems. Trees regenerating after coppicing may look similar.) |
| H | ▽ | Herbs (Non-woody erect plants including most flowers and grasses.) |
| M | | Bryoids (Cushion plants including mosses, liverworts, lichens, fungi.) |

**SIZE**

| | | Trees | Shrubs | Herbs | Bryoids |
|---|---|---|---|---|---|
| t | Tall | >25m | 2-8m | >2m | – |
| m | Medium | 10-25m | 0.5-2m | 0.5-2m | >10cm |
| l | Low | 8-10m | <50cm | <50cm | <10cm |

**FUNCTION OF LEAVES** (Seasonal and other activities)

| | | |
|---|---|---|
| d | | deciduous |
| e | | evergreen |
| s | | succulent |
| l | | leafless |

**LEAF SHAPE AND SIZE**

| | | |
|---|---|---|
| n | ⬭ | needle or spine (e.g. pine, spruce, firs or gorse.) |
| g | ◊ | graminoid (e.g. grasses, sedges.) |
| m | ◇ | medium or small (e.g. privet leaf or smaller.) |
| b | △ | broad or large (larger than privet leaf.) |
| v | ∨ | compound leaf (e.g. ash or horse chestnut ) |
| q | ◯ | thalloid (lichens, thalloid liverworts) |

**LEAF TEXTURE**

| | | |
|---|---|---|
| f | | filmy (very thin-leaved water plants) |
| z | | membranous (most leaves – e.g. oak, ash, beech, elm.) |
| x | | sclerophyllous (hard and tough – evergreen needles, heather, holly.) |
| k | | succulent (fleshy – stonecrops, house-leek.) |

**COVERAGE**

| | |
|---|---|
| b | barren or very sparse |
| i | discontinuous (<60%) |
| p | in tufts or groups |
| c | continuous (>60%) |

**Fig. 4.12** Symbols proposed by Dansereau for structural records of vegetation. Letters and symbols may be used together or on their own

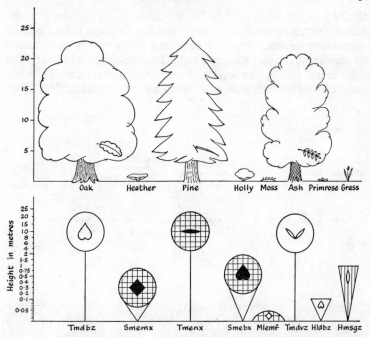

**Fig. 4.13** Structural records of the same vegetation using realistic diagrams (above) and symbols (below). Note how height is shown

### Belt transect

A belt transect (Fig. 4.14*b*) is a strip, usually ½ m or 1 m wide, marked out by setting out a second line parallel to the original transect line. The vegetation between the lines is carefully mapped in the field notebook, completing one square at a time. Continuous patches of the same plant can be mapped by some shading code, otherwise individual plants should be represented by suitable code letters. The belt transect is often used to amplify the picture given by a profile transect and when so used the width of the belt can sometimes be reduced without detracting from the value of the record.

### Belt transect histogram

In this method the floristic and faunistic composition of each ½ m or 1 m square is recorded, the percentage of the square occupied or covered by each species being estimated as accurately as possible. For some purposes a *presence or absence record*, listing species present in each square without estimating frequency, is sufficiently meaningful when looked at in conjunction with habitat factors such as tide or salinity (see Fig. 12.2, p. 174). Identification in each square must be accurate because transitions from one species of a genus to another may otherwise be missed. Various habitat factor values are also determined, not necessarily in every square. The field notebook should be prepared with vertical rulings (for level, habitat factors and species) and horizontal rulings for each station at which records are taken (Fig. 4.15).

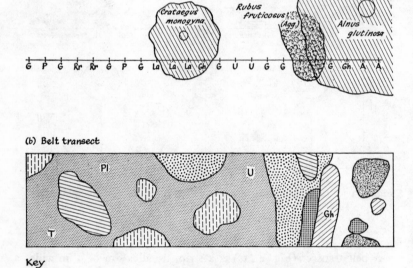

(a) Line transect

(b) Belt transect

Key

A Arctium minus

Gh Glechoma hederacea

G Unidentified grasses

La Lamium album

Rr Ranunculus repens

Pl Plantago major

U Urtica dioica

P Potentilla anserina

T Taraxacum officinale

**Fig. 4.14** (a) Line Transect; vertical intercepts of shrub and tree canopies shown and herbs recorded every 15 cm. (b) Belt Transect; individual plants recorded by letter symbols, continuous masses by shading or colouring

In practice it is often found that no purpose is served by recording every square along a transect and that a useful record may be made by examining only every third square.

The results are plotted as a chart on graph paper, the percentage

## STATION

| (50 cm quadrats at 1 metre intervals) | 1 | 2 | 3 | 4 | 5 | 6 | 7 | 8 | 9 | 10 | 11 | 12 | 13 | 14 | 15 | 16 | 17 | 18 | 19 | 20 |
|---|---|---|---|---|---|---|---|---|---|---|---|---|---|---|---|---|---|---|---|---|
| Level (change from previous station in cm) | 0 | -30 | -30 | -23 | -27 | -33 | -42 | -37 | -43 | -53 | -120 | +56 | -26 | 0 | -10 | -99 | +165 | +86 | +89 | |
| Soil depth in cm | 20.5 | 24 | 25.5 | 56 | 60 | 18 | 16.5 | 47 | 61 | 23 | 15.5 | 51 | 63.5 | 66 | 29.5 | 28 | 77.5 | 96.5 | 74 | |
| Light meter reading | 11 | 11.25 | 11.5 | 12 | 12 | 12 | 12.5 | 12.75 | 13.25 | 13.0 | 12.0 | 13.0 | 13.9 | 11.5 | 10.5 | 12.0 | 12.5 | 12.5 | 12.5 | |
| Soil pH | 4.8 | 4.7 | | 4.8 | | 4.7 | | | | | 5.1 | 5.0 | 5.9 | | | | 5.0 | 4.8 | 4.8 | |
| Soil moisture (per cent) | 32.1 | 30.6 | | 28.4 | | 18.7 | | | | | 41.6 | 37.1 | 41.0 | | | | 28.2 | 24.7 | 32.3 | |

## SPECIES

| (percentage cover) | 1 | 2 | 3 | 4 | 5 | 6 | 7 | 8 | 9 | 10 | 11 | 12 | 13 | 14 | 15 | 16 | 17 | 18 | 19 | 20 |
|---|---|---|---|---|---|---|---|---|---|---|---|---|---|---|---|---|---|---|---|---|
| Galeobdolon luteum | 50 | 30 | 25 | 10 | | | 10 | 30 | 40 | 5 | 50 | 5 | 15 | 60 | 10 | 15 | 45 | 15 | | |
| Anemone nemorosa | 5 | | 10 | 15 | 20 | | 10 | | | | | | | | | | | | | |
| Endymion non-scriptus | 5 | 5 | | | 5 | 10 | 25 | | | | | | | | | | | | | |
| Veronica sp. | 20 | 5 | 15 | 5 | | | | | | | | | | 5 | | | | | | |
| Lonicera periclymanum | 15 | | | | | | | 35 | 10 | | | | | | | | | | | |
| Ferns | 5 | 40 | | | | | | | | | | | | | | 10 | 5 | | 5 | |
| Mercurialis perennis | | 15 | | 50 | 30 | 30 | | 35 | 30 | | | | | | | | | | | |
| Bare ground | | 5 | 30 | 15 | | 60 | 65 | 25 | | 20 | 75 | 20 | | 15 | | 50 | 80 | 55 | 80 | |
| Circaea lutetiana | | | 20 | | 45 | | 10 | 5 | | | | | 5 | | 15 | | | | | |
| Rubus sp. | | | | | | | | 50 | | | | | 35 | 50 | 15 | | | | | |
| Chrysosplenium oppositifolium | | | | | | | | | | | 20 | 25 | | | 5 | | | | | |
| Urtica dioica | | | | | | | | | | | | | 60 | | | | | | | |
| Oxalis acetosella | | | | | | | | | | | | | | | | 20 | 30 | | | |

NOTES
Map ref. of wood — 985193  (O.S. Map No. 144)

Date. 28th. May 1957

Transect runs due North-South

Station 5 against hazel tree

Station 11 falls further 25 cm after 22 cm horizontal, then rises. Soil sample at 11.5.

Stations 12-15. Clay soil. Thinner canopy.

Station 16 preceded by fall of 127 cm from Stn. 15. Stream 30 cm wide. Stream bottom hard.

**Fig. 4.15** Page of field notebook—record for Belt Transect Histogram. (Note: soil depth was determined by probing with a steel rod)

coverage or actual numbers being plotted by blacking in a solid block (Fig. 4.16). Level, habitat factors, floristic and faunistic compositions can be compared vertically. Using this method transitions become more obvious and their relationships to habitat factors are suggested.

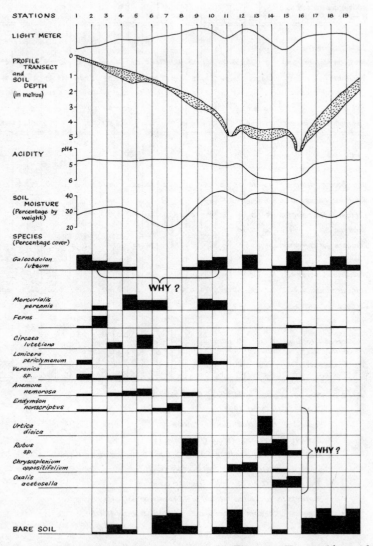

**Fig. 4.16** Belt transect histogram prepared for Fig. 4.15. Two problems of plant distribution are indicated. The profile and habitat factor records suggest solutions which must be checked by further investigation

## Quadrat frames

A robustly constructed metal or wooden frame is often used to mark out the areas investigated when making a belt transect histogram (Figs. 4.17 and 4.18a). This can be used with a 'comb' (having 10 teeth and used at 10 points on the frame) to sample 100 equally spaced points; this automatically gives an estimate of percentage.

**Fig. 4.17**   Using a quadrat frame to sample vegetation

In some habitats inspection may show that a belt transect histogram as described above may be less profitable than making a large number of records at each of rather fewer parts of the transect. In this case the quadrat frame can be used for *random* or *systematic sampling*. If the frame is thrown to fall at 'random' the personal element may enter and better results are in fact obtained when the frame is placed as indicated by *random number tables** using a numbered grid on the map of the area. Systematic sampling at even intervals is often used for showing up suspected variation.

Because of the difficulty in obtaining a truly random distribution it may be more useful to increase the number of samples but reduce the size of the quadrat frame and possibly use it without a comb. The

* Fisher and Yates (1951): *Statistical Tables for Biological, Agricultural and Medical Research.*   London, Oliver & Boyd.

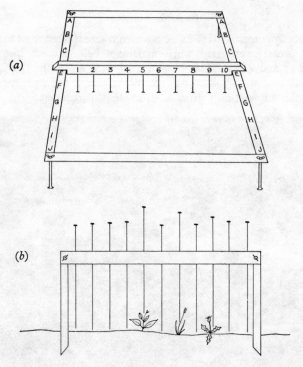

**Fig. 4.18**   (a) Quadrat frame and 'comb'. (b) Point Quadrat

best size will naturally depend on the nature of the vegetation, but it should be remembered that a small quadrat has more edge in proportion to area, and it is at the edge that the observer is continually having to decide what to include or exclude, depending on whether rooted species or cover is the main test.

An indication of when further sampling is becoming unprofitable is obtained by graphing the number of species against the area sampled (Fig. 4.19).

Instead of using a frame and comb, a *point quadrat* can be obtained from a free-standing comb with loose pins whose points are lowered on to the vegetation (Figs. 4.18b and 4.20). Each piece of vegetation on the way to the ground is recorded, and also the bare ground if hit. Some workers prefer to use only one pin, recording hits as it is lowered through each hole in turn. When recording the frequency of a particular species the result may be expressed as a percentage of total hits; other species need not necessarily be recorded by name.

A point-quadrat frame may be modified for use as a *topograph* (Fig. 4.21). A clip-board with graph paper is attached to the frame so

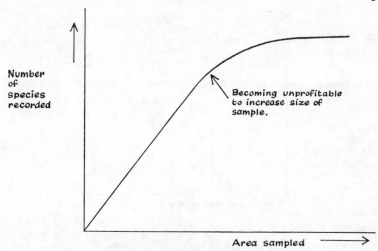

**Fig. 4.19** Relationship between number of species and area sampled. Sampling area is increased either by using larger quadrat frames, or by using larger numbers of the same size frame

**Fig. 4.20** Point quadrat in use on sand dunes

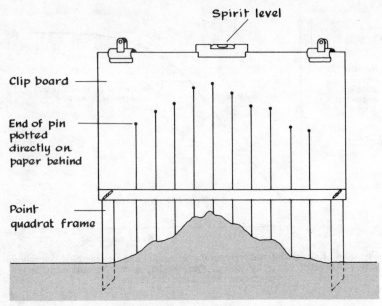

Fig. 4.21    Point quadrat modified for use as a topogragh

that the upper ends of the pins may be plotted directly and quickly to record the profile of the ground beneath; this is particularly useful when recording plants associated with tussocks.

### Permanent quadrat

Changes in flora and fauna in successive seasons through the year, and possibly in successive years, can be followed by setting out a permanent transect line or, more usually, a permanent quadrat. Such a quadrat is a rectangular area of ground, clearly marked out by boundary stakes or other permanent marks. Records of plant and animal life and of habitat factors are made at regular intervals. The size of such a permanent quadrat may vary from squares with sides no more than a few centimetres long (in the case of submerged plates for tracing algal succession) to squares or rectangles with sides as much as 10 metres in length. With even the largest quadrat it is possible to map the vegetation inside it by direct measurement. A map prepared in this way should also bear an accurate indication of positions used for soil tests, temperature readings, light estimations and so on. Individual organisms, locations of tests and other relevant information should be marked on the map by clear letters, symbols and shading as in a belt transect.

## Other records

Maps, transects and quadrats may not contain all the information needed for a full understanding of plant and animal distribution in an area. To complete the background picture a *consolidated species list* (C.S.L.) with *frequency assessment* should be prepared. The C.S.L. is a list of all the organisms recorded for the area. The organisms should be roughly divided under group headings – in animals by phylum or habitat (soil, leaf gall, etc.), and in plants it may be useful to combine taxonomic groups (ferns, bryophytes) with life-forms (trees, shrubs, herbs).

However the list is compiled, a frequency assessment should be made for each species. This assessment is usually denoted by one of the letters D (dominant), A (abundant), C (common), F (frequent), O (occasional), R (rare). In certain instances the prefixes Co-, Sub-, or V (very) may be added; these should be used sparingly. Letter assessments have little meaning on their own except to the person using them *at or near* the time of use. The term **dominant,** for instance, may be used in two different ways; it may refer to the species providing the highest density and cover, or may be used to indicate the species that exerts the greatest influence on other members of the community, and is least influenced by them. For records to be useful to others, or to the originator after some time has elapsed, all these terms must be clearly defined. For instance, Crisp and Southward (1958) *J. Mar. Biol. Ass. U.K.*, **37**, p. 160, defining the abundance of limpets on a rocky shore, use the letter system as follows:

A   Over 50 per m² or more than 50% of population limpets at certain levels
C   10 to 50 per m², 10 to 50% at certain levels
F   1 to 10 per m², to 10% at certain levels
O   Less than 1 per m² on average, less than 1% of population
R   Only a few found in 30 minutes searching

However, for barnacles which are much smaller, the terms have a very different meaning:

A   More than 1 per cm²; rocks well covered
C   0.1 to 1.0 per cm²; up to one-third of rock space covered
F   0.01 to 0.1 per cm²; individuals never more than 10 cm apart
O   0.001 to 0.01 per cm²; few within 10 cm of each other
R   Less than 1 per m²; only a few found in 30 minutes searching

Thus symbols must be defined to suit the species, though a number of species of similar size and habit may be grouped using the same notation. Broadly speaking, plant abundance can be defined by reference to its growth form, though some subdivision will be required;

grasses with a high number of plants per unit area will probably need a different notation from other herbs. In particular, individual rosette plants cover a relatively large area of ground, so that the impression of abundance is given by only a few plants.

Nothing has been said so far about **seasonal changes**; just as presence or absence records may be made along a transect line, so they should be made in relation to time. These seasonal records should include, besides presence or absence of annual plants, such details as the time of first appearance of leaves, flowers, fruits, leaf-fall in perennial plants, and emergence from hibernation, mating, appearance of young or spawn in animals. Records on individual organisms should include full descriptions of change in form and habit; the winter forms of many common plants are imperfectly known.

Existing lines like the line of a stream can often replace transects as a basis for sampling. When recording on a stream, suitable sites are selected (Chapter 11) and numbered as stations for survey. At such stations it is often helpful to have a sketch profile showing depth and width of stream, height and slope of banks and overshadowing trees. An estimate of water speed should be included in records of the habitat factors (discussed in Chapter 6).

## Productivity and Energy Flow

To study energy flow through a food chain it is obviously necessary to determine the productivity at each trophic level, together with losses from respiration and wastage (see p. 232). Useful information for most purposes may be gained from determining the biomass incorporated into only the aerial parts of the plant community. This is known as *net primary aerial production* and involves harvesting or estimating the amounts of leaf, stem, seed and associated organs produced over the period studied. For grassland, the samples can be harvested by hand-cutting all herbs growing through fixed quadrat frames fitted with stout wire mesh. The grass is trimmed to the level of the wire at the start, and harvesting merely involves trimming back to the same level and measuring the weight of the dried product. The time interval is obviously critical – frequent harvesting will correspond to heavy grazing, and inhibit fruit and seed formation. Seasonal effects can also distort results – productivity can be estimated fully only if the whole growth cycle is studied. The number of species compared with the number of individuals in a sample gives an index of the *diversity*; in a trophic level, diversity tends to confer stability and adaptability to change. In this kind of investigation it is obviously necessary to exclude grazing animals from the sample area during the period of study, and similar considerations apply to other plant communities.

*Secondary* ( = *consumer*) *production* at each level is determined through population estimates coupled with analyses of size, weight and breeding condition to determine age distribution, birth, growth and death rates. Food consumption and assimilation has to be determined both by analysing stomach contents of trapped animals and by laboratory trials to determine amounts of food taken and the weight of faeces produced. Respiration rates are again best estimated in the laboratory.

Outlines of the main methods for assessing primary productivity in different communities are made in the I.B.P. Handbooks on *Methods for Measuring Primary Production* (Blackwell Scientific Publications).

## Animal numbers

The profile, belt transect and quadrat methods are primarily designed for investigating plant life, and are generally useful only in the case of sessile or fairly slow-moving animals, such as barnacles and molluscs. This type of recording, however, is specially valuable for demonstrating clearly correlations between particular animal and plant species, as between molluscs and algae on the seashore. A *total* count of the population in a habitat is known as a **census** and is only occasionally possible with fast-moving animals. A direct count can be made of bats in their roost, deer, territorial birds in woodland, and sometimes small mammals when a habitat such as a hayrick is being taken apart in an enclosure. Mobility and secretive habits mean that most other animal populations must be *estimated* indirectly by counting **samples**.

Small, fairly evenly distributed animals, such as plankton in a lake or pond, can be estimated by taking samples and counting the numbers in a given volume or area. If enough samples have been taken, a reasonably reliable figure for the overall population density can be calculated. The sampling method and the number and position of samples should be decided only after careful consideration of the habits and distribution of the species concerned. This is necessary because agile species may often avoid entering sample tubes, nets or traps, so leading to underestimation of the population size. Alternatively the animals may be distributed very unevenly, so that too few samples could give a quite misleading measure of the population. Sometimes a knowledge of behaviour may be helpful, as in the case of earthworms known to be attracted by dung. A trench is excavated around an area not more than 1 metre in radius. Then dung is placed in the centre of the isolated area and examined for worms at intervals of a few days. On moorlands, where populations are sparse, this may be the best way of estimating earthworms.

The capture–recapture technique is useful where an animal can be distinctly labelled with a ring, metal tag or spot of paint. A number of animals are caught, marked and then released. Later a second sample is taken and the number of recaptured marked individuals is noted. The size of the total population can now be calculated.

$$\text{Total population} = \frac{\text{Number in 1st sample} \times \text{Number in 2nd sample}}{\text{Number of marked individuals recaptured}}$$

This method suffers from several drawbacks. It is difficult to apply lasting marks to many species, and often the labels render the animals more conspicuous to predators or hinder their movement. Random mixing of the population, which in theory must occur between the first and second sampling, is seldom complete, and may be influenced by births, deaths or migration into or out of the area. Birds and small mammals, to which the method is often applied, tend with experience to become trap-shy or trap-addicted, so that the proportion of marked individuals found in the second sample may be quite misleading. A compensating advantage of this type of sampling is that in cases where numbered rings or colour codes can be used, a great deal of information about development, life span and territory can be gained.

Small insects like ants or fleas may be marked by placing them in a tube containing a powdered aniline dye. The tube is shaken and they are then released. Insects caught in the next sample are all killed and placed on filter paper. If a drop of suitable organic solvent is placed on each insect, fragments of dye trapped in crevices on recaptured animals are dissolved and cause conspicuous colour marks on the filter paper.

Numbers can also be estimated, though with a much smaller degree of accuracy, by counting traces such as food remains, tracks, dung or burrows. Such methods apply particularly to birds and mammals with secretive habits. For instance, the number of open burrows can be used as a measure of the size of a badger population. The ratio of badgers to open holes can be observed in detail in a small district and the result extended to a count of burrows over a much larger area.

Many mammals, including otters, mice, hares and foxes, can be studied indirectly by counts of their faecal pellets or droppings. Useful records of many birds can be made by noting the frequency of specific calls or songs, especially in the case of nocturnal or woodland species. Again, the stomach contents and pellets of predators can give an indication of the abundance of various species of prey, as for example mouse and vole remains in owl pellets. It should be noted, however, that predators do not sample at random, and that changes in the frequency of different prey species may simply reflect a change in the hunting preferences of the predator.

Populations of some species, particularly small active forms, cannot easily be estimated in absolute terms, but can usefully be compared in a relative way. The animals are collected in a standard manner, with a net or by hand, for a given time, and the catches from different places are compared. This technique has been used successfully with leeches, molluscs, triclad flatworms and leafhoppers. The main source of error when collecting by hand or net is the rapidly improving efficiency of the collector as he gains experience in the early stages. Several preliminary trials are therefore necessary in order to standardise the collector. If automatic methods are employed for collecting, these also should be standardized. An example of this is a bundle of crushed earthworms left for a standard period in a stream to attract planarians.

With small mammals, *line-trapping* can be used to show where they are most frequent within a habitat. The basis of this method is to select a line which passes through a varied part of the habitat, and set traps close to 'stations' established at regular intervals. Thus a trap-line 50 metres long might contain 50 traps, 5 being set in a rough circle around each of ten positions. By using this technique to take a single sample only, the problem of trap-addiction is avoided.

Since the density and distribution of animal populations are so difficult to record accurately, it is generally advisable to estimate in more than one way. Thus a vole population might be studied by the capture–recapture method, by counts of skulls in owl pellets, and by noting the quantity of faecal pellets dropped in a given area. Furthermore, it is important to consider the life-history, particularly in invertebrates, because a census may be possible only for the adults, and the population density of these may depend upon factors acting on the larvae.

Animal numbers are never static; this is why their study is known as population *dynamics*. It is also why population measurements should be repeated over a considerable period including times of growth and decline, particularly in relation to the factors which cause these changes. Fluctuations may be seasonal, long term (building up to a 'crash') or a combination of both; the source of such fluctuations is likely to differ with the habitat. Where a population is building up to a crash as often happens with the vole *Microtus*, the animals often move into abnormal habitats. Some habitats are invaded annually, there being no winter survival – as happens when insects invade fields after overwintering in hedgerows. Following such invasions populations grow exponentially until the limitations of physical conditions or competition for food and space set an equilibrium level about which the population oscillates (Fig. 4.22). Changes in predator populations tend to follow those of their prey, but it must be remembered that few predators depend on only one species as a food source.

**Population income** is made up of birth-rate and immigration; **population expenditure** is by death and emigration. Determination of the age structure of the population, from live or killed samples, can

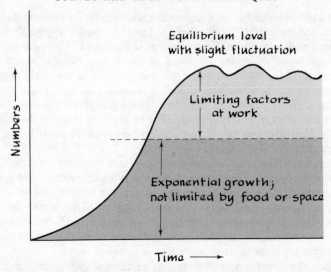

**Fig. 4.22** The pattern of population growth following invasion of a new habitat. Growth is exponential until limiting factors start to act and establish an equilibrium level

form the basis of a *life-table* giving the number in each age-group, and hence the possibility of assessing birth and survival rates from one age group to the next. Censuses or mark-and-release estimates in two successive years give an index of change in the population as a whole and make possible a prediction of future numbers. This will of course only be valid if birth and survival rates are unaffected by changes in habitat resources or activity of predators. If the population is stable and not changing significantly from one year to the next (and this observation may be the starting point of some investigations), study of *income* will provide information about expenditure, and *vice versa*.

# 5 Using Quantitative Records

Numerical records must always be matched to their intended use. It is therefore necessary to think and plan ahead, trying to anticipate which kinds of information will be most useful; the best recording method is then usually obvious. Failure to consider statistical aspects and control experiments in advance may result in an experiment from which no valid conclusions can be derived. A simple investigation relating plant frequency to habitat factors is the belt transect histogram (p. 60); the controls in this case are provided by other parts of the transect.

## Frequency distribution

In many cases, quantitative records concern measurements of size, weight and number relating to individuals or parts of individuals. Examples include leaf length, seed weight and flower number in plants, and body weight, limb bone lengths and numbers of eggs laid in animals. Because no two individuals are exactly alike, any description of a particular **population** of organisms must be based on a fairly large **sample** (number of records). If these records are plotted on graph

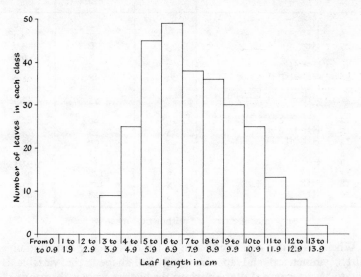

**Fig. 5.1** Histogram recording variation in apple leaves. $\frac{5}{6}$ of all measurements lie between 4 and 11 cm

paper it is easy to see which values describe most of the population. Thus in Fig. 5.1 apple leaves are mainly between 4 and 11 cm long. In order to determine this **central tendency** of the sample, the variable being investigated is divided into convenient **classes,** and the data is sorted and assigned to the different groups to form a **frequency distribution.**

Where the characteristic varies **discontinuously** by whole numbers, each number group may be used as a frequency class. However, if there are many such number groups, or if the characteristic varies **continuously** (as in measurements of weight or height), it is best to assign the records to classes and determine the frequency distribution by constructing a histogram as in Fig. 5.1.

The number of flowers per inflorescence in *Anthriscus sylvestris* (cow parsley) provides an example of discontinuous variation in which the frequency distribution is determined by 'scoring' the inflorescences in each number group. For convenience it is usual to score with vertical strokes crossed off by horizontal strokes to complete each group of five. If the scoring is done neatly and evenly it is not necessary to make a histogram to see the shape of the resulting frequency distribution.

TABLE 5.1 — FLOWER NUMBER VARIATION IN *Anthriscus*

| Flower Number $x$ | Frequency Score | Frequency Totals $f$ | Cumulative Frequencies (see p. 84) Totals | % |
|---|---|---|---|---|
| 5 | II | 2 | 2 | 1.6 |
| 6 | I | 1 | 3 | 2.3 |
| 7 | III | 3 | 6 | 4.7 |
| 8 | II | 2 | 8 | 6.2 |
| 9 | ++++ III | 8 | 16 | 12.5 |
| 10 | ++++ ++++ II | 12 | 28 | 21.9 |
| 11 | ++++ ++++ ++++ ++++ ++++ II | 27 M | 55 | 42.9 |
| 12 | ++++ ++++ ++++ ++++ ++++ | 25 * | 80 | 62.5 |
| 13 | ++++ ++++ ++++ | 15 | 95 | 74.1 |
| 14 | ++++ ++++ I | 11 | 106 | 83 |
| 15 | ++++ | 5 | 111 | 86.5 |
| 16 | ++++ I | 6 | 117 | 91.5 |
| 17 | III | 3 | 120 | 98.8 |
| 18 | ++++ | 5 | 125 | 97.6 |
| 19 | III | 3 | 128 | 100 |

M = modal class    * arithmetical mean $\simeq$ 12.2

Where, as in this example (Table 5.1), the frequency distribution is fairly symmetrical and approximately bell-shaped, the variable is said to have a **normal distribution**; the biggest score is the **mode** of the sample and that number-group may be used as the one most

typical of the population. In this example, a slightly bigger score in
the higher flower-number classes has resulted in the **average** or
**arithmetical mean** of the population being in a different class from
the mode; with a larger sample this difference might disappear.

Where *continuously* varying characteristics are grouped to form a
histogram, we distinguish a **modal class** instead of a mode, and
there may possibly be no single measurement that is exactly repeated
within this class. When grouping into classes, considerable care needs
to be taken over **class intervals** and **boundaries**. The class interval
should be of such a size that the entire range is divided into a convenient
number of classes. A convention must be established as to which class
a record on the boundary is to be assigned. For instance, in height
measurements with a class interval of 3 cm, a class may have boundaries
at 9 cm and at 12 cm and the class may include the lower boundary
but not the upper; if measurements are to the nearest millimetre, the
class can be defined as running from 9.0 to 11.9 cm inclusive. To calculate Standard Deviation (see below), the **mid-class number** (10.4 in
the class just defined) should be used in the $x$ column (Table 5.2).

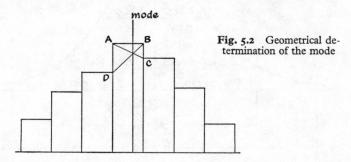

**Fig. 5.2** Geometrical determination of the mode

With a histogram, an approximation of the mode can be made by
dividing the modal class geometrically, as in Fig. 5.2.

If AB is the modal class, the estimated mode runs through the
intersection of AC and BD.

Modal values obtained in this way can provide an indication of
differences between populations; such differences may be genetic
(inheritable) or environmental – the result of variation in the habitat.
Many common weeds have a wide range from mildly xerophytic to
nearly hydrophytic conditions and populations from the two extremes
are likely to show significant differences in structure and dimensions.

## Standard deviation

Only a small proportion of the sample represented in a frequency
distribution corresponds to the arithmetical mean, the remainder

being dispersed on either side. A useful measure of this dispersion is provided by the **standard deviation** (*S*) of the curve. It is the distance from the arithmetical mean (*never* from the mode or from the median or mid-point) to the steepest part of the curve and is the **root mean square** of all deviations from the mean. It can be calculated quite simply from a scored frequency distribution, without having to draw the normal curve which best fits the results. Standard deviation can be used as a rough measure of probability because roughly two-thirds of the population falls in the area of the curve within one standard deviation from the mean, roughly 95% falls within two standard

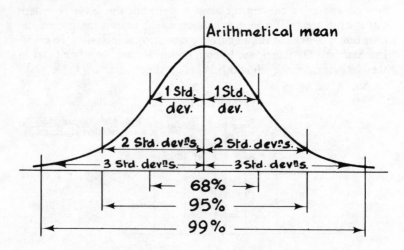

**Fig. 5.3**   Standard deviation and the normal curve (see text)

deviations, and well over 99% falls within three standard deviations.

Calculation of the standard deviation is simple provided that the results are set out systematically. The mean is rarely a whole number, so to avoid having to square a lot of decimal figures it is usual to use a whole number near the mean and make a correction afterwards.

In our example of *Anthriscus* flower numbers, we take the modal class (11) as our working mean from which to calculate deviations. It will be seen in Table 5.2 that the first two columns represent our original score, the third column $(x-a)$ represents the deviation and the remaining two columns are obtained in turn by multiplying together

TABLE 5.2

| Flowers per inflores-cence x | Frequency f | Deviation from working mean $(a = 11)$ $x-a$ | $f(x-a)$ | $f(x-a)^2$ |
|---|---|---|---|---|
| 5 | 2 | −6 | −12 | 72 |
| 6 | 1 | −5 | −5 | 25 |
| 7 | 3 | −4 | −12 | 48 |
| 8 | 2 | −3 | −6 | 18 |
| 9 | 8 | −2 | −16 | 32 |
| 10 | 12 | −1 | −12 | 12 |
| 11 | 27 | 0 | 0 | 0 |
| 12 | 25 | +1 | 25 | 25 |
| 13 | 15 | +2 | 30 | 60 |
| 14 | 11 | +3 | 33 | 99 |
| 15 | 5 | +4 | 20 | 80 |
| 16 | 6 | +5 | 30 | 150 |
| 17 | 3 | +6 | 18 | 108 |
| 18 | 5 | +7 | 35 | 245 |
| 19 | 3 | +8 | 24 | 192 |
| TOTALS | $\sum f = 128$ | — | $\sum f(x-a) = +152$ | $\sum f(x-a)^2 = 1166$ |

corresponding figures in the preceding two columns. The $f$, $f(x-a)$ and $f(x-a)^2$ columns are totalled at the bottom; these totals are referred to in the formulae as $\sum f$, $\sum f(x-a)$ and $\sum f(x-a)^2$ respectively. ($\sum f$ is also referred to as $N$, as we have in connection with correlation, p. 81.)

With these results we correct for the mean as follows:

$$\text{Correction} = \frac{\sum f(x-a)}{\sum f} = \frac{152}{128} = +1.188$$

Thus    true mean $\bar{x}$ = working mean + correction

$$= 11 + 1.188 = 12.188$$

Standard Deviation is given by the formula:

$$S = \sqrt{\frac{\sum f(x-a)^2}{\sum f} - \left(\frac{\sum f(x-a)}{\sum f}\right)^2}$$

$$= \sqrt{\frac{1166}{128} - (1.188)^2}$$

$$= \sqrt{7.701} = 2.775$$

Over 95% of records from the same population can be expected to lie within 2 Standard Deviations of the mean; i.e. on average, less than 5% of inflorescences will have fewer than 6 or more than 18 flowers.

From the formula it is obvious that the Standard Deviation is to a considerable extent determined by the size of the sample ($\sum f$). The Standard Deviation of a large sample will be much smaller than that of a small sample. The quickest way of showing this is to take the frequency curve we have been using, then double the score for each flower number and see how the shape of the curve is changed from

to

Fig. 5.4

The *standard deviation of the sample* (S) is nearest in value to the *standard deviation of the whole population* (σ) when the sample is large.

## Standard Error of the Mean

With a wide frequency distribution some variation in the *mean* of successive samples can be expected. An estimate of the range of this variation is provided by the Standard Error of the Mean. This is calculated from the Standard Deviation:

$$\text{S.E.} = \frac{S}{\sqrt{\sum f}}$$

In our example, then:

$$\text{S.E.} = \frac{2.775}{\sqrt{128}} = \pm 0.2431$$

For a 95% **range of confidence** the mean

$$\simeq \bar{x} \pm 2[\text{S.E.}] = 12.188 \pm 0.486$$

In other words the mean will lie between 11.7 and 12.7 for 95% of all similar samples.

If a similar investigation made on a different population of *Anthriscus* showed a quite different and separate range for the mean, this would suggest that the difference might be due to some significant cause which might be genetic (inheritable) or phenotypic (resulting from the influence of habitat factors). In fact it was found that similar populations from Cheltenham, Dorset and Devon showed good agreement for the position of the mean.

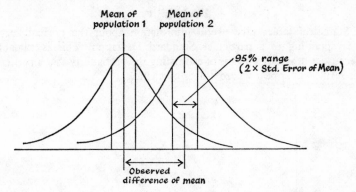

**Fig. 5.5** Diagram of two frequency distributions showing significant difference between the means

The inflorescence of *Anthriscus* is a compound umbel and this investigation involved counting the number of flowers in terminal umbels from different plants. A similar investigation carried out on

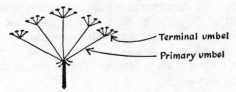

**Fig. 5.6** Inflorescence of *Anthriscus sylvestris*

the *primary* umbels showed a significant difference in populations quite close to one another but in different conditions of shade and soil.

## Standard Error of the Difference

The difference between two populations is investigated by determining the Standard Error of the Difference, given by:

$$\text{S.E. of Difference} = \sqrt{\frac{S_1{}^2}{\sum f_1} + \frac{S_2{}^2}{\sum f_2}}$$

where $S_1$ and $S_2$ are the standard deviations of the samples from the two populations, and $\sum f_1$ and $\sum f_2$ are the numbers in the populations.

As before, the 95% range of confidence is given by doubling the S.E. of the Difference. If the observed difference lies outside this range, there is a high probability that there is some underlying cause other than random sampling.

## Probability

Statistical tables give useful information about the normal curve corresponding to a particular Standard Deviation. This is valuable because it provides a means of estimating the probability that a particular result will appear where it does by chance.

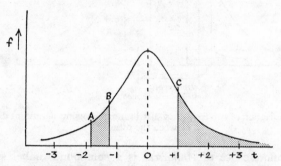

**Fig. 5.7**    Areas under the normal curve (see text)

An approximation of the probability of a record coming in a limited part of the frequency distribution is provided by the ratio of the area under this part of the curve to the total area under the curve. The total area under the curve represents **total probability** ($P = 1$). The curve flattens towards its limits but for most practical purposes it may be regarded as ending at three Standard Deviations on either side of the arithmetical mean.

TABLE 5.3

*Areas under the normal curve* (one tail only)

| $t$ | 0.00 | 0.25 | 0.50 | 0.75 | 1.00 | 1.25 | 1.50 |
|---|---|---|---|---|---|---|---|
| $P$ | 0.500 | 0.401 | 0.309 | 0.227 | 0.159 | 0.106 | 0.067 |
| $t$ | 1.75 | 2.00 | 2.25 | 2.50 | 2.75 | 3.00 | |
| $P$ | 0.040 | 0.023 | 0.012 | 0.006 | 0.003 | 0.001 | |

Figures from Fisher & Yates: *Statistical Tables for Biological, Agricultural and Medical Research,* Oliver & Boyd Ltd., Edinburgh, by permission of authors and publishers

Reference to Table 5.3 gives the probability that any individual record in the sample will fall outside any particular multiple ($t$) of Standard Deviations from the mean. The value given for probability refers to only one tail of the curve, that is to say only in the half of the curve on one side of the mean. Thus in Fig. 5.7, the probability of a result coming in the part of the curve to the right of C, one Standard Deviation from the mean ($t = 1.0$), is $P = 0.159$ (or 15.9%). The probability of a record coming *within* one Standard Deviation of the mean in the right-hand tail is the difference between the probability for the whole tail ($t=0$; $P=0.5$) and the last figure; i.e.

$$P = 0.5 - 0.159 = 0.341$$

The probability of coming within one Standard Deviation of the mean in both tails will be the sum of the values for each tail:

$$P = 0.341 + 0.341 = 0.682 \text{ (or 68.2\%)}$$

In a similar way, the probability of a result coming between A and B is obtained by determining the areas to the left of each, and subtracting the value for A from that for B.

A single result coming in any particular position under the curve provides little information about the population in general; such a result is said to have little **significance**. Where the result is repeated its significance increases. We have seen that the probability of a result coming in the part of the curve to the right of C (in Fig. 5.7) is 15.9% ($P=0.159$). The probability of two consecutive results coming in this region is obtained by multiplying the two probabilities together. Thus $P=(0.159)^2=0.026=2.6\%$.

Where the alternatives are clear cut, it is not necessary to refer to tables to determine probability. For instance, theoretically there are two equal chances of being right when predicting the sex of an unborn baby. This means that we can say there is a probability of 1 in 2 or $\frac{1}{2}$. In fact this is not quite true so it is safer to continue this discussion by referring to the non-biological example of coin tossing.

If a coin is tossed five times and shows heads each time, this is not nearly as significant as if it were tossed fifty times and still kept showing heads. In the latter case it would be more than reasonable to suppose that something was causing the result to be biased. The probability of each individual coin showing heads is $\frac{1}{2}$ (there being only two alternative possibilities). The probability for five tossings is found by multiplying the probabilities for individual coins together. This gives $\frac{1}{2} \cdot \frac{1}{2} \cdot \frac{1}{2} \cdot \frac{1}{2} \cdot \frac{1}{2}$ or $(\frac{1}{2})^5$ which is $\frac{1}{32}$. This means that this result can be expected at least every 32nd time. For fifty tossings, the probability is $(\frac{1}{2})^{50}$ which means that such a result is likely to occur by chance less than once in a thousand million million times!

In practice, results are rarely as clear cut as this and the problem might be to assess whether 26 heads out of 50 tossings can be regarded as significant. It cannot, because it is quite probable that this result would be obtained simply by chance. However, 2600 out of 5000 is less likely by chance and is correspondingly more significant; 26 000 out of 50 000 (a deviation of 1000 from the expected result) is even more unlikely. The probabilities, and hence significance, of such results can be obtained from statistical tables using a mathematical formula which takes into account the number of samples as well as both the observed and the expected results. This *Chi-squared* $(\chi^2)$ test, which is described in the books recommended for further reading, may be applied to many different kinds of biological investigation including ones where, as in breeding experiments, there are more than two different expected results. In every case, the records must be extensive if they are to reveal significance in a small deviation from the expected result.

Sometimes a frequency distribution curve for a particular variable character is found to be asymmetrical or **skew** instead of normal in shape. This may indicate selection pressure due to some habitat factor at work on one side of the curve. An example of a situation in which a skew distribution could result from man's activities is where intensive fishing with nets is carried out. The mesh size of the net selects the majority of fish above a certain age and girth.

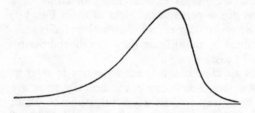

**Fig. 5.8**　A skew curve

In much the same way, animals like the whale and herring which feed by filtering plankton could upset the size distribution by selecting plankton above a certain size.

Not all selection is **directional**; in some cases it may be **stabilizing** and tend to reduce the spread of a frequency distribution, whilst in other cases it may be **disruptive** and split the distribution into two or more groups. These different kinds of natural selection, working on the variation present in the population, are a major cause of evolution.

## Correlation

Where a relationship between two variables is suspected, it may be investigated by preparing **scatter diagrams**. Thus, in preparing such a diagram to investigate the relationship between an organism and a particular habitat factor, the frequency of the organism is plotted against the corresponding factor value for a fairly large number of samples. The result is a scatter of points which may or may not reveal a **correlation** between the two. Such a correlation is revealed when the scatter of points shows a definite trend in one direction or the other. If an increase in factor value is accompanied by an increase in population, the correlation is said to be positive. If the population decreases as factor value rises, the correlation is said to be negative.

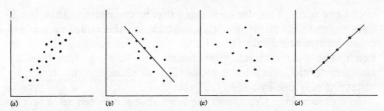

**Fig. 5.9** Scatter diagrams: (a) strong positive correlation; (b) strong negative correlation with line fitted by eye; (c) absence of correlation; (d) perfect positive correlation (r=1)

Very often a straight line can be drawn by eye to fit the data; it is usually possible to see directly whether this gives a fair indication of the general trend. Its reliability can be assessed by determining the **correlation coefficient** (r) and then consulting tables relating it to the number of samples (see below). A curved line cannot be assessed so easily. In general it should be noted that a wide scatter to which a line cannot easily be fitted means a low correlation coefficient (approaching r=0), whilst a set of values on or close to the line has a high correlation (approaching r=+1 or r=-1).

For 95% probability of correlation (P=0.05), minimum values of r are (approximately):

| Number of samples | N | 10 | 15 | 20 | 25 |
|---|---|---|---|---|---|
| Correlation coefficient | r | 0.63 | 0.51 | 0.44 | 0.4 |

Figures from Fisher and Yates *Statistical Tables*.

In the scatter diagram (Fig. 5.10) relating the caddis fly *Agapetus* with rate of water flow, it is possible to draw a line but the results are not meaningful because the points are too scattered, and there are

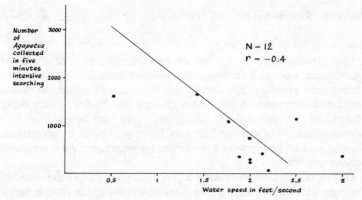

**Fig. 5.10** Relationship between *Agapetus* numbers and water speed (see text)

only 12 of them. This does not mean that *no* correlation exists, but it does suggest that if there is a correlation, (*a*) the sampling method could be improved and (*b*) a larger number of samples should be taken. The same correlation coefficient (*r* = −0.4), derived from 25 samples, would indicate the probability of obtaining this result by chance at only one in 20 (5%; *P* = 0.05).

In contrast, Fig. 5.11 shows the relationship between an aspect of flea behaviour and the time after emergence from the cocoon. The positive correlation is immediately obvious, and the grouping of the points makes line fitting relatively easy. The correlation coefficient (*r* = 0.95) indicates a very strong correlation which is more than 99.9% significant (*P* = 0.001). In other words, there is a probability of less than one in 1000 that these results were due to chance alone.

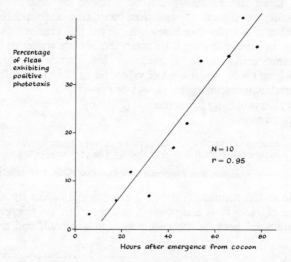

**Fig. 5.11** Relationship between flea behaviour and time of emergence from cocoon (see text)

Considerable care is needed in the interpretation of apparent correlations, and such investigations should normally be made only after other evidence has suggested a relationship. An apparent correlation between two different organisms may be due to each being affected by the same habitat factor, and there may be no *direct* link between them. The *indirect* link may be very difficult to determine; there is said to be a strong correlation between the divorce rate and banana imports!

Choice of units when making a graph directly affects the slope of the regression line so that a poor correlation may appear much better than it really is.

Many kinds of situation can be investigated by this method; here are some suggestions:

| rate of water flow | and | numbers of plants or animals |
| rate of flow | and | growth form (e.g. percentage of aerial or floating leaves in *Sagittaria*) |
| light intensity | and | frequency (plants in woodland, animals under stones and in litter, etc.) |
| soil moisture | and | plant form (leaf size and succulence, length of internodes) |
| moisture | and | animal numbers (animals in soil or rotting bark) |
| air humidity | and | bats, insects and spiders in caves |
| soil acidity (pH) | and | plant or bacterial frequency |
| salinity of soil water | and | cell sap concentration or growth form (on salt marshes) |
| temperature | and | animal numbers (many animals, from free-living flatworms to ectoparasites, have been alleged to show positive or negative thermotaxis) |

### Calculating the Correlation Coefficient

The correlation coefficient for a linear (straight-line) relationship can easily be calculated if results are set out systematically as for the relationship between flea behaviour and time (Fig. 5.11 and Table 5.4).

TABLE 5.4

| 1<br>Hours after emergence of fleas<br>$x$ | 2<br>Percentage showing positive phototaxis<br>$y$ | 3<br>Deviation of $x$ from mean ($\bar{x} = 44$)<br>$d_x$ | 4<br>Deviation of $y$ from mean ($\bar{y} = 22$)<br>$d_y$ | 5<br>$d_x^2$ | 6<br>$d_y^2$ | 7<br>$d_x.d_y$ |
|---|---|---|---|---|---|---|
| 6 | 3 | − 38 | − 19 | 1444 | 361 | 722 |
| 18 | 6 | − 26 | − 16 | 676 | 256 | 416 |
| 24 | 12 | − 20 | − 10 | 400 | 100 | 200 |
| 32 | 7 | − 12 | − 15 | 144 | 225 | 180 |
| 42 | 17 | − 2 | − 5 | 4 | 25 | 10 |
| 48 | 22 | + 4 | 0 | 16 | 0 | 0 |
| 54 | 35 | + 10 | + 13 | 100 | 169 | 130 |
| 66 | 36 | + 22 | + 14 | 484 | 196 | 308 |
| 72 | 44 | + 28 | + 22 | 784 | 484 | 616 |
| 78 | 38 | + 34 | + 16 | 1156 | 256 | 544 |
| TOTALS:<br>$\sum x = 440$ | $\sum y = 220$ | | | $\sum d_x^2 = 5208$ | $\sum d_y^2 = 2072$ | $\sum d_x.d_y = 3126$ |

(NOTE: With a wide scatter, some of the values for $d_x d_y$ may be negative and must be *subtracted*.)

The two variables $x$ and $y$ are set out in columns 1 and 2. Each of these columns is totalled and divided by the number of samples ($N = 10$) to determine its arithmetical mean. Thus:

for column 1:
$$\bar{x} = \frac{\sum x}{N} = \frac{440}{10} = 44$$

for column 2:
$$\bar{y} = \frac{\sum y}{N} = \frac{220}{10} = 22$$

Column 3 contains the deviations of samples in column 1 from the arithmetical mean. Thus the first figure in column 3 is given by $d_x = 6 - \bar{x} = 6 - 44 = -38$. All the other figures in the column are obtained in the same way. Column 4 contains the deviations of samples in column 2 from $\bar{y}$. The figures in columns 5, 6 and 7 are calculated from the corresponding figures in columns 3 and 4.

Covariance of $x$ and $y$

$$S_{xy} = \frac{\sum d_x . d_y}{N} = \frac{3126}{10} = 312.6$$

Standard Deviation of $x$

$$S_x = \sqrt{\frac{\sum d_x{}^2}{N}} = \sqrt{\frac{5208}{10}} = 22.8$$

Standard Deviation of $y$

$$S_y = \sqrt{\frac{\sum d_y{}^2}{N}} = \sqrt{\frac{2072}{10}} = 14.4$$

Correlation coefficient

$$r = \frac{\text{Co-variance of } x \text{ and } y}{\text{Product of Standard Deviations of } x \text{ and } y}$$

$$= \frac{S_{xy}}{S_x . S_y} = \frac{312.6}{22.8 \times 14.4} = 0.95$$

## Cumulative Frequency

For certain purposes it may be useful to record *cumulative frequency*. In this, the record for each class includes the accumulated frequencies of preceding classes. Thus in Table 5.1 (p. 72), cumulative frequencies read '2, 3, 6, 8, 16, . . .' compared with class frequencies '2, 1, 3, 2, 8, . . .'. Plotting cumulative frequencies for a normal distribution gives an 'S' shape curve on ordinary graph paper (Fig. 5.12), but a straight line on 'probability paper'. The latter gives a simple way of confirming that a distribution is normal. A table of cumulative frequencies may be

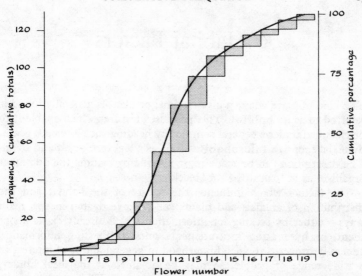

**Fig. 5.12** Cumulative frequency curve based on Table 5.1. Both cumulative totals and cumulative percentages are indicated; the increment for each flower class is shaded

used to determine quickly how many of a sample lie above, below or between certain values, as when analysing the age, weight or height distribution in productivity or population–dynamics studies. Cumulative frequencies may also be expressed as percentages.

# 6 Habitat Factors

The kind of place where a given animal or plant is generally found is referred to as its **habitat**. This possesses features which enable it to be classified under a general community heading such as wood, heath, dune, lake, or as a **microhabitat** such as a log, bird's nest and so on. The description can be made more specific by naming the dominant organism as in 'oak wood' or 'blackbird's nest'.

The factors which influence the nature of the habitat and the distribution of animals and plants can be investigated in two main ways – either by seeking for information which already exists about them, or by making measurements oneself. Existing information generally concerns the neighbourhood in general and is of interest to a wide range of people, so may be obtained at the local library, meteorological station, university or field centre. When making one's own measurements, it is easier to work with other interested groups such as a Geography Department in a school or university.

To understand the way in which habitat factors work, it is often necessary to draw on laboratory studies in anatomy or physiology. Thus, during respiration and photosynthesis, animals and plants exchange gases with the atmosphere through moist surfaces, and if the surrounding air is unsaturated some of this moisture will evaporate. Most land organisms reduce water loss by enclosing their moist surfaces and presenting to the atmosphere a relatively impermeable cuticle with few openings for gaseous exchange. Some are less waterproof than others so that in experimental conditions they show different rates of water loss which are seen to show a close relationship to their natural distribution in the field. Amongst isopods (woodlice and related forms), *Porcellio* is confined to the moist air of leaf litter whereas the more resistant *Armadillidium* is able to wander further afield. On the seashore, isopods like *Ligia* and *Idotea* are normally restricted to the damp air under rocks and seaweeds. However, rocks exposed to full sunlight at low tide may become dangerously warm. Experiment has shown that these isopods can maintain a lower temperature than their surroundings by emerging at intervals into the drier air and losing heat as a result of evaporation from the body surface (Fig. 6.1). Thus features like cuticular water loss which at first sight appear restrictive may sometimes be essential for survival. With plants, light uptake for photosynthesis is always accompanied by heat absorption, and water loss by transpiration plays an important part in preventing overheating of the leaf and 'leaf-burn'.

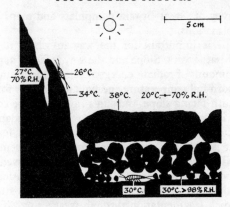

**Fig. 6.1** Vertical section of the base of a red sandstone cliff and shingle inhabited by *Ligia oceanica* (diagrammatic) to show the microclimatic conditions and internal temperatures of the animals at *c.* 1400 hours G.M.T. in August 1951. [From Edney (1953), *J. exp. Biol.*, **30**.]

Habitat factors are grouped under different headings for convenience, but will often be found to be interdependent.

## Topographic factors

The shape of the ground is related to the underlying rock structure. Ordnance Survey maps provide topographical information, and the

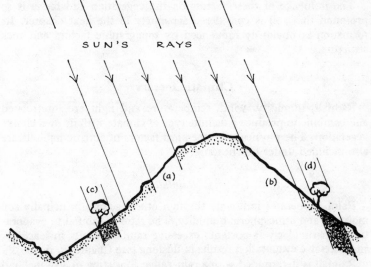

**Fig. 6.2** Effect of aspect and slope on spread of sun's rays and shadow length. (*a, c*) South facing slope; (*b, d*) North slope where vegetation is less vigorous in growth

Geological Survey of Great Britain pamphlets and maps give details
of geological formations.

Ground shape is important for the way in which it can modify
many other habitat factors. Slope and the way it faces (**aspect**) partly
determine the amount of radiant energy reaching the ground from the
sun, affecting both soil temperature and light availability for photosynthesis. On a south-facing slope, the sun's rays are more concentrated
so that the soil is warmer and plant growth more abundant than on a
north-facing slope, where more moisture and shade loving plants
grow. Gradient is important in relation to water movement and its
effect on the soil; there is a tendency for soil to be removed from steep
slopes and deposited lower down. The effects of erosion are more
pronounced if the slope cuts across the underlying rock strata because
there is more rock fragmentation through weathering. If the rock is
shaly, the eroded fragments may give rise to an unstable scree on which
plants can barely become established. Owing to surface run-off, soil
on steep slopes generally contains less water than in more level places.

Hills and high ground are more exposed to wind and generally
undergo greater fluctuations in temperature and humidity than valleys.
In valleys the **water-table** (level below which the soil is waterlogged)
is often high compared with that on higher ground.

## Edaphic (Soil) factors

The influence of these factors on the vegetation and fauna is so
profound that soil is considered separately in the next chapter. Its
formation is obviously influenced by topographic factors and rock
structure.

## Climatic factors

Rainfall, humidity, wind, temperature and light are interrelated
and combine to produce a definite type of climate with its own climax
vegetation. The physical and chemical factors of aquatic habitats are
also included under this heading.

### 1. *Rainfall*

Rainfall operates indirectly through other factors, principally soil
moisture and atmospheric humidity. The ratio of rainfall to evaporation is particularly important; excessive rainfall results in leaching,
and excessive evaporation results in flushing (see Chapter 7, p. 106).

Rainfall is determined with a rain-gauge consisting of a funnel and
collecting cylinder, and the measurements are expressed as centimetres
per day or week. These results should be recorded graphically on a

**time-chart** which can be related to other time-charts or records concerning changes in vegetation, soil fauna and so on. The gauge is normally sunk into the ground so that its funnel is flush with the surface as this reduces evaporation losses; it should be visited every two or three days, and more frequently during very wet weather.

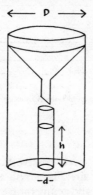

**Fig. 6.3**
Rain gauge (see text)

When improvising such a gauge using a cylinder of uniform diameter, the rainfall in centimetres is $\dfrac{d^2}{D^2} \cdot h$ where $d$ and $D$ are the diameters of cylinder and funnel respectively, and $h$ is the height in centimetres of water in the cylinder. Thus if $D = 10$ cm and $d = 2.5$ cm, the actual rainfall is $\frac{1}{16}$ the observed height $h$.

## 2. *Humidity*

Atmospheric humidity is important for its effect on the rate of water loss from both animals and plants; this rate is lowest when the air is most humid. In caves, woodlands and other sheltered places, the air is moister than elsewhere. Many animals and plants can live only in such places, whilst others can limit their water loss and therefore have a wider range.

Except in microhabitats, humidity is best measured using a **whirling hygrometer**. This consists of wet and dry bulb thermometers arranged in a frame which can be rotated like the moving part of a football fan's rattle (Fig. 6.4). The hygrometer is whirled until the two temperature readings are constant. Then, the **relative humidity** is quickly determined from psychometric tables purchased with the hygrometer. This value gives the percentage humidity relative to saturated air at the same temperature and gives no indication of the absolute water content of the air. For many purposes humidity is better expressed as **saturation deficit** which gives a quantitative indication of the amount of water that the air can still take up. The relationship between relative humidity and saturation deficit is indicated in Fig. 6.5. Because **saturation vapour pressure** (vapour pressure at 100% Relative Humidity) rises with the temperature it

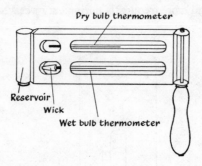

Fig. 6.4
Whirling hygrometer

will be seen that, for any given relative humidity, warm air can take up more water than cool air. The graph also shows the relationship between vapour pressure and temperature at various humidities.

In microhabitats the best method is probably to use small paper or hair hygrometers in which a pointer is moved as the paper or hair changes in length. These need to be standardized in a variety of humidities which have been accurately determined by other methods (see Fig. 6.6). Where an accurate chemical balance is immediately available, the increase in weight when dry paper strips are placed in a microhabitat can indicate the humidity. The percentage increase should be compared with the increases when whole sheets of similar paper are placed in known humidities. (The standardization results should be recorded on a graph; values obtained during experiment may then be extrapolated from this.) Another useful method is to time the colour change of cobalt chloride papers from a standard blue to standard pink. This method is often used to determine the rate of transpiration of a leaf; most biological laboratories will have a supply of both cobalt chloride paper and the standard colour papers. Other cobalt salts show similar colour change. The Tintometer Ltd., Salisbury, produce a humidity test kit using cobalt thiocyanate test papers and a set of coloured glass standards for direct comparison.

Any attempt to use the wet and dry bulb method with thermometers or thermistors (see below) in a confined space should be avoided, because this is likely to change the humidity one is trying to measure.

A useful comparative method for measuring the total evaporating power of air is the **atmometer** which measures the actual evaporation from a moist surface. The apparatus used is similar to a potometer, but has a porous thimble or other evaporating surface replacing the plant shoot. Any kind of potometer may be adapted in this way (Fig. 6.7 (a) and (b)). A simple atmometer may be made from a piece of capillary tube (Fig. 6.7 (c)) by grinding one end flat on an oil stone and fitting a rubber tap washer to support a disc of filter paper, cut out with a cork borer. This end of the tube should be bent through a right angle so that both the longer part of the tube and the evaporating surface are horizontal. The tube is filled with water and a moist disc of filter paper

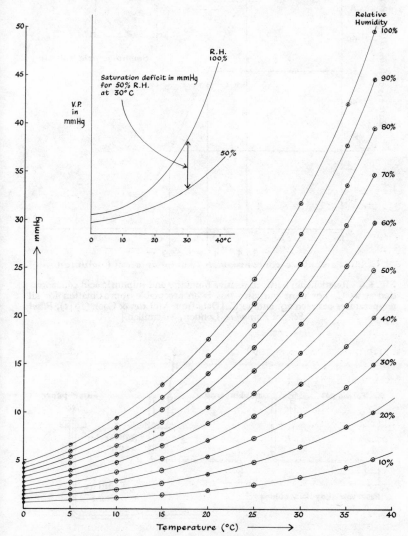

**Fig. 6.5** Graph for determining the saturation deficit for a given temperature and relative humidity. Inset shows method of determination

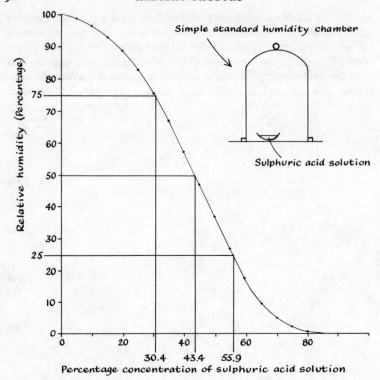

**Fig. 6.6** Relationship between relative humidity and sulphuric acid concentration at 25°C. For most purposes this is an acceptable approximation for all temperatures between 20° and 30°C. [Data from McLean & Cook (1941), *Plant Science Formulae*, London, Macmillan.]

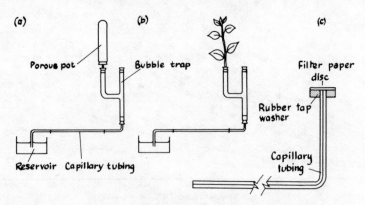

**Fig. 6.7** (*a*) and (*b*) Darwin potometer in use as an atmometer and as a potometer. (*c*) Simple capillary tube atmometer

applied. As the water evaporates air is drawn in and its speed of movement measured. Because gusts of wind increase evaporation, the average
of several readings should be taken after the apparatus has first had
time to reach a temperature equilibrium with its surroundings. This
apparatus can be used to find out how transpiration rate is modified by
environment. Useful comparisons might be made between the inside
and outside of a wood, at different heights, in the open and by a hedge
or other windbreak, and on north and south sides of hedges on sunny
and overcast days. It can prove instructive to follow such comparisons
with potometer experiments.

Where the average evaporation over a longer period needs to be
determined, a different type of atmometer due to James may be used
(Fig. 6.8). Here the upper end of the tube entering the porous thimble

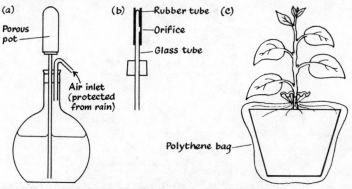

**Fig. 6.8** (*a*) 'James' atmometer' showing detail (*b*) of non-return valve.
(*c*) Potted plant for use as 'phytometer'

has a simple valve which prevents backflow of water if the apparatus
is exposed to rainfall. A 250 cm$^3$ flask acts as a reservoir, and there is a
downward-facing air inlet. Transpiration rates over similar periods can
be measured using potted plants as 'phytometers' and measuring
weight loss per hour in different situations. The pot must be enclosed
in plastic material and sealed with plasticine to prevent water loss
directly from the soil surface.

Humidity is affected by the plants and animals themselves and by
the presence or absence of wind. Woodlice in a dry atmosphere tend
to congregate so that the humidity in their immediate vicinity rises;
massing of plant leaves in a rosette similarly raises humidity and
therefore limits water loss.

### 3. *Air and and water movement*

Wind is most important for its effects on transpiration and on
evaporation from moist surfaces generally; the resulting coldness of

soils in exposed places affects the character of the vegetation pro-
foundly (see Chapter 10). On cliff tops where there is a strong prevailing wind the trees may show asymmetric growth; this, and the
uniform height of woodland trees, is largely due to death of the most
exposed shoots as a result of excessive transpiration.

As well as for mechanical damage and leaf-fall, wind is also important in the dispersal of certain seeds and fruits, some spiders and
small insects such as aphids.

Anemometer readings obtained from meteorological stations are of
limited value because wind speed tends to vary according to its direction and to the lie of the land. Simple wind-gauges can be improvised
(Fig. 6.9) and are most useful for comparative studies such as com-

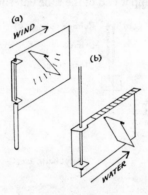

**Fig. 6.9**  Deflection flow-meters. (*a*) Wind gauge (to be viewed from side).
(*b*) Water-current gauge (to be viewed from above)

paring wind speeds inside and outside a wood, or at different heights
above ground.

Water flow affects the distribution of both animals and plants;
fast-moving water is well aerated but is inhabited only by organisms
whose structural or behavioural adaptations enable them to maintain
their position. Still water tends to contain less oxygen.

The simplest method of measuring water speed is to time the
movement of a floating object over a measured distance. A light
object on the surface may be affected by wind, so unless there is no
wind the object used should be weighted so that it barely breaks the
surface; an orange or partly filled specimen tube serves well for this.

Flow-meters are needed to measure the flow at different depths;
the simplest types are:

(*a*) PITOT-TUBE AND MANOMETER (Fig. 6.10). This is constructed
so that the manometer will be well above the surface of the deepest
water investigated. The sensitivity of this device depends on the

density of the manometer liquid, light liquids giving greatest sensitivity; the difference in height in the two arms of the manometer is recorded. Care must be taken to align the instrument with the flow, either by attaching a tape as 'weather-vane' or by rotating the apparatus about its vertical axis until the maximum reading is obtained. The

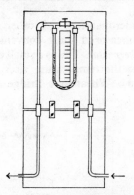

**Fig. 6.10** Pitot tube and manometer. Note the tap to by-pass the manometer when not in use. The sections of the mounting board may be bolted together as shown, or hinged

manometer method is not very good for slow moving water. The instrument can be calibrated in stretches of water where the flow is seen to be sufficiently uniform for determination by the floating-object method.

(*b*) DEFLECTION METER. Here water flow deflects a hanging, weighted vane from the vertical; the instrument can be made as sensitive as desired by adjusting length and breadth of the vane, and the weight of its lower end. It is aligned either by using a simple streamer or by building the meter on to an enlarged weather-vane. The instrument is calibrated as in (*a*) above.

Another important aspect is exposure to wave action or other types of water erosion; there is no simple way of assessing this.

## 4. *Temperature*

Most living organisms have an optimum temperature range, because low temperatures slow down the important metabolic processes associated with respiration, growth and reproduction, whilst high temperatures cause irreversible changes which may lead to death. Wherever possible, temperatures should be recorded on time charts showing the day to day fluctuation in maximum and minimum tem-

peratures as recorded on a Six's thermometer. Such a chart can then
be related to the time scale of seasonal changes in plant and animal
activities. For animals, single temperature readings in microhabitats
may be more meaningful than in plants, though here, too, a series of
readings is to be preferred. Such results may reveal the extent to
which the microhabitat shelters inhabitants from more extreme
temperatures outside.

Temperatures in microhabitats and other relatively inaccessible
places can often be measured most conveniently with an electric
thermometer using a *thermistor*. This is a small capsule whose electrical
resistance decreases as the temperature rises, and is connected by
low-resistance leads to a Wheatstone Bridge circuit (Fig. 6.11). By

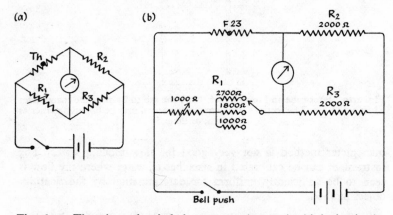

**Fig. 6.11** Thermistor electrical thermometer (see text): (*a*) basic circuit;
(*b*) a circuit suitable for determining temperatures up to 40°C using a Standard
Telephones & Cables F23 thermistor

direct marking of a scale behind the control knob of the variable
resistance $R_1$, the instrument may be calibrated in a water-bath
against a good mercury thermometer. Resistances $R_2$ and $R_3$ have the
same value so that $R_1$ is equal to the thermistor resistance when the
galvanometer shows a zero reading. By using a series of shunts, $R_1$ can
be arranged in several narrow ranges to give readings of greater
accuracy. When calibrating, all measurements should be made *using
the actual leads* which are to be used in the field. The instrument will
need to be recalibrated if the leads are altered in any way. Only high-
stability resistances should be used.

The high specific heat of water means that water temperatures vary
much less than air temperatures. This is particularly true where
there is a large volume of standing water as in a pond or lake. In
shallow side streams the temperature may vary much more because

the greater surface-to-volume ratio favours more rapid heating in sunlight and cooling in shade. Further downstream, time charts show that peak temperatures occur later in the afternoon or evening, when the warm headwaters flow down.

Use of an ordinary mercury thermometer in water is limited by a tendency for the reading to change while the thermometer is brought from deep water to the surface. This tendency is reduced if the thermometer bulb is first coated with an insulating layer of wax. A well protected maximum/minimum thermometer left in the water works well for any chosen time interval. Thermistor thermometers will give direct readings at any depth but must be effectively waterproofed; twin PVC-covered cable is particularly suitable and can be marked at intervals for direct measurement of depth.

In soil, as in air, a time chart of temperatures at different depths is more valuable than any single measurement. Surface temperature can be read directly, whilst for depths of more than a few inches a permanent shaft of drain-piping may be used so that a wax-insulated thermometer can be brought to the surface for periodic examination.

## 5. Light

Light is important because it provides the energy used by green plants in photosynthesis; all animals depend on complex foods produced in plants by this process. Light is also important in relation to animal behaviour – many organisms, like woodlice and soil animals, find optimum conditions by moving towards or away from the light. Other organisms show responses directed towards light of a particular strength so that in water there may be daily vertical migrations.

Hours of daylight are given in most pocket diaries, but hours of sunlight need to be specially recorded.

The simplest way to determine light intensity is to use an ordinary photographic light meter. The meter will need recalibrating if readings are to be given in standard units (lux; lumen/m$^2$); it is generally more useful to use meter readings unaltered for comparing habitat light intensity with that under open sky at the same time. Indirect readings using light reflected from a white surface are better than direct readings; any value recorded should be the average of several readings.

This method can be adapted for underwater light measurements using encapsulated photoelectric devices attached to a suitable handle. Alternatively, less expensive unmounted selenium or cadmium sulphide cells may be sealed into perspex containers, though care is needed because perspex cement contains materials which may cause cells to deteriorate in time. Leads from a selenium cell may be connected directly to a sensitive moving coil 0 – 0.5 mA milliammeter. Cadmium sulphide cells and phototransistors require batteries, and also resistors to bring the current into the range required by the meter; the range of readings can be extended to meet all likely needs by introducing a series of shunts. Other photoconductive cells include lead sulphide cells which have their peak response in the infra-red and so may be used as infra-red detectors.

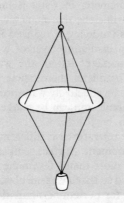

**Fig. 6.12**　Secchi disk

The amount of light available under water is related to the presence of suspended matter and plankton. Comparative measurements of turbidity from these causes can be obtained using a Secchi disk. Here, the depth at which a horizontal white disk just disappears from view is recorded. A plastic dinner plate, suitably supported and weighted, makes a satisfactory disk for this purpose (Fig. 6.12).

### 6. Water composition

All living organisms are descended from saltwater ancestors, and this is reflected by the complex homeostatic mechanisms which still maintain salty body fluids in freshwater and terrestrial animals. Colonization of fresh water and land can only be achieved by organisms which can control their water relations; organisms which can adapt to a changing salinity are said to be **euryhaline**, whilst those with only a narrow range of tolerance – most marine, freshwater, and terrestrial animals – are **stenohaline**. Salinity determines the range of most animals and plants found in the boundary areas between sea and freshwater (estuaries, salt-marshes), and between sea and land (the seashore).

Sea water is fairly constant in composition and provides the mineral salt, oxygen, and carbon dioxide requirements of its inhabitants. Fresh water is less constant, and contains only small traces of mineral salts. Of these, calcium appears essential for many freshwater animals and enables some estuarine species to tolerate greater dilution. Running water is generally richest in oxygen, whilst standing water contains most carbon dioxide; however, photosynthetic activities of water plants may produce great changes in oxygen level during a 24 hour period in still water.

SAMPLING. In most cases a screw-top jar is adequate for obtaining specimens down to arm's-length depth. For deeper samples a weighted bottle with ground glass stopper may be used. This is lowered gently by means of a string tied firmly round the neck, and about 10 cm higher, to the stopper. When the bottle has reached the required depth, the stopper is removed by jerking the line. With either method little air is likely to dissolve in the water entering. However, for very accurate sampling special arrangements are made to ensure that the water collected has not come into contact with the air originally in the sampler. Contaminated water is drawn off by suction or driven into an accessory vessel by the inflow of water (see Appendix C).

SALINITY. A useful approximation for salinity is given by titrating 10 cm$^3$ samples of sea water with a solution containing 27.25 g of silver nitrate per litre from a burette. This concentration is such that the volume (in cm$^3$) of the silver nitrate required roughly equals the salinity of the sample (in g NaCl per litre). Potassium dichromate is used as indicator. The indicator turns red, but the end point is given by the first slight soiling of the precipitate. A quick alternative method is to determine the temperature and density of the sample and use a conversion graph to estimate total chlorides (Dowdeswell (1959), *Practical Animal Ecology*, p. 54, London, Methuen).

OTHER DISSOLVED SUBSTANCES. Detailed chemical methods for other dissolved substances are beyond the scope of this book but some useful summaries are included in Appendix C. A full account may be found in *Water Analysis for Limnologists* by F. J. H. Mackereth (Freshwater Biological Association Scientific Publication No. 21). Colorimetry using a B.D.H. Lovibond Nessleriser and prepared solutions provides a quick means of determining dissolved oxygen; the same apparatus with different colour standards can be used for a variety of tests including pH determinations (see also Chapter 7, p. 122). A drawback of colorimetry is that the apparent colour of the artificial standard varies with the source of light; strong diffuse daylight reflected from a white card should always be used.

## Biotic factors

These factors stem from the activities of living plants or animals and operate in a variety of ways, few of which can be assessed quantitatively.

Plants are particularly important, because they affect both the food and habitat resources available to animals. Many herbivorous animals are restricted in diet and cannot survive in the absence of food plants in suitable condition. *Aphis fabae*, the bean and beet aphid, reproduces so rapidly that competition for the same food can lead to relative shortage; quality of the food is important also, and this aphid shows a marked preference for juvenile and senescent leaves, producing fewer offspring on mature ones. In many other animals, including man, population explosions may produce serious food shortage as a result of competition between members of the same species or with members of another species. This is very obvious in the case of pests competing with humans for food crops and other products, but may also happen in the wild. Most plants support a number of different herbivorous **primary consumers**, and if one of these has a particularly successful year, the others will go short, and so will their predators and parasites (see Fig. 9.6). Again, food availability for both herbivore and carnivore is affected by long-term as well as seasonal changes in the vegetation.

Populations are often regulated by density *dependent* factors. Overcrowding may result in less food per individual and a reduced reproductive rate; parasites establish themselves quickly and reduce numbers directly or by spreading disease. Myxomatosis was spread by the rabbit flea; when rabbits became rare, they were less likely to acquire virus-carrying fleas from other rabbits and so survived and reproduced. Crowding often disturbs animal behaviour, reducing fertility in some and provoking migration in others. Interactions occur between trophic levels; predator and prey act as density dependent regulators on each other – there is less predation when prey populations are small. Within a species, aggression and the establishment of territories can ensure the effective exploitation of habitat resources for survival and breeding.

One organism may provide or create microhabitats for others. Trees and shrubs provide nesting sites for a variety of birds, each nest provides a habitat for bird lice or fleas, and some of these parasites are found on the bird itself, which in turn provides a home for many internal parasites. Such living habitats are found everywhere, tall plants dominating the community and providing shelter for less resistant species which in turn may provide food and shelter for a miscellany of other animals and plants.

## The human factor, and pollution

Man, through his industrial and agricultural activities, is the most important biotic factor affecting the environment. In grassland, grazing and trampling (sheep paths, near gates) cause changes in flora and fauna, whilst arable crops provide ideal conditions for population explosions of both weeds and animal pests. Weed and pest success depends largely on dispersal mechanisms and rates of growth. In general there is tremendous wastage, only very small proportions of seeds or fertilized eggs ever reaching maturity. What causes this wastage ? Why is there less wastage among certain weeds and pests of a particular crop ? This needs to be worked out for each organism; important reasons include competition for light, nutrients, breeding and feeding territories, as well as predation and parasitism. Farming methods tend to upset the balance between predator and prey; **biological control** is achieved by finding a way of adjusting the balance to suit the farmer. Insecticides and herbicides may have quite the wrong effect because they leave no reserve populations of predators to deal with new infestations.

Fire is caused by man as a regular agricultural practice, or by sheer carelessness. However induced, it can have a profound influence on both flora and fauna; a particularly damaging effect is when the soil itself catches fire – peat may burn for years, and the whole ecosystem can be destroyed. Controlled burning of heather and gorse on moorland helps to halt natural succession and maintain the general character of the habitat; it also provides opportunities for more grass and other herbs to grow and support grazing animals or game birds. Forest fires destroy hundreds of years of tree growth and kill many animals. Stubble burning is rather different, for it is merely removing the dead remains of a crop of a single plant species; what animals are associated with corn-crops, and how are they affected by this procedure?

Industrial activities and some agricultural practices are particularly important in causing **pollution** of soil, water, and atmosphere. Pollution is the production or release of substances which alter the environment and make it less favourable to the animals and plants that live there.

Some polluting agents can be broken down by living organisms and are therefore short-lived; they are said to be biodegradable. Other pollutants are stable and last longer. Domestic sewage is biodegradable and modern sewage treatment aims at decomposing and recycling organic materials in a way that avoids adverse effects on the environment. *Small* quantities of untreated sewage discharged into the sea or rivers may be broken down by bacteria and recycled naturally without affecting other organisms. Greater quantities, however, result in increased numbers of bacteria whose demand for oxygen is so enormous that larger organisms can no longer survive. An indication of bacterial activity, and hence of organic pollution, is given by determining **Biochemical Oxygen Demand (B.O.D.)** – the oxygen consumption over a five day period (see pp. 161 and 238).

Non-degradable materials include poisons from metals, radioactive waste, and chemicals used in agriculture such as fertilizers and organo-chlorine pesticides (e.g. DDT and Dieldrin). All tend to accumulate in the environment and can affect it adversely. The widespread practice of using more than recommended amounts of pesticides and fertilizers is particularly harmful; the excess is carried away with drainage water and affects the life of streams and rivers. Some pesticides kill as a result of accumulating in the tissues of the consumer; the rate at which they

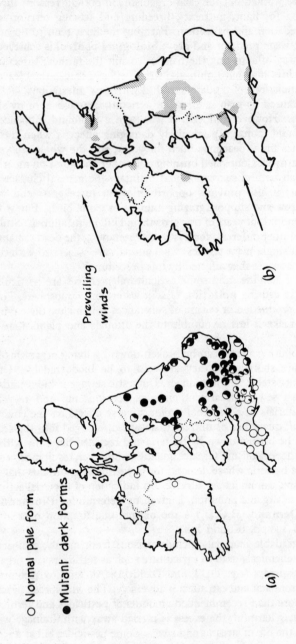

**Fig. 6.13** Maps showing (a) the frequency of light and dark forms of the peppered moth, *Biston betularia*, and (b) the most densely populated areas and prevailing westerly winds. Dark forms are found in industrial areas and to the east of them. (Map (a) after Ford, E. B., 1971, *Ecological Genetics*, 3rd edn. Map 7 Chapman and Hall.)

○ Normal pale form
● Mutant dark forms

Prevailing winds

collect increases in successive links of a food chain as each co. turn takes in material that has already been concentrated many t. In this way, use of herbicides and insecticides to check weeds and insects may have unforeseen indirect effects – the unusually high mortality of foxes during the winter 1959–1960, originally thought to be due to disease, was later shown to be the result of eating birds poisoned by seed dressings. Since then, many examples affecting other animals have been reported. Although many seed-eating birds have died of poisoning, their numbers may have actually increased as a result of the increased death rate of predators.

Combinations of pollutants occasionally react to produce additional and more harmful substances. Nitrogen oxides and hydrocarbons passed into the atmosphere with the exhaust of aero- and motor car engines combine in sunlight to produce the mixture known as 'photochemical smog' – toxic to both animals and plants. Lead from exhaust fumes also gives rise for concern; in the U.S.A., food plants grown near busy roads have been found to contain unacceptable levels of lead. Carbon monoxide pollution may reach significant levels in city centres but the effect on individuals is still less than the carbon monoxide poisoning they could suffer from heavy smoking.

Other important atmospheric pollutants are sulphur dioxide and smoke. Lichens are particularly susceptible to sulphur dioxide and can be used as indicators of pollution. Which lichens are most sensitive? What parts of town are most affected? Can you locate the main source of pollution? Wiping the leaves of evergreen plants with clean white tissues will give a measure of smoke pollution in different areas. The soot darkened trees and buildings of our towns have resulted in dark coloured mutants of many common moths having a selective advantage over lighter forms which are more readily detected and eaten by birds (Fig. 6.13). Are the proportions of normal and dark forms changing following the establishment of smokeless zones?

There is not space to mention in detail the many other forms of pollution. The barrenness of mine waste heaps is often due to the presence of toxic minerals. Drilling for oil in the North Sea increases the possibility of oil pollution again affecting our shores and seabirds, and the use of river water to remove heat and other industrial products may result in profound changes in the organisms found there. Noise, fast-moving vehicles and litter including broken glass and sharp-edged cans, all add to the hazards of animal life.

**Conservation** is an important human activity which aims to halt or reverse undesirable changes resulting from pollution, erosion, or even natural plant and animal succession. It includes any measure which tends to conserve an existing pattern in nature; burning moorland shrubs, hedging and ditching, and forestry management, are all well-established procedures which have this effect. Planned conservation of rare habitats reduces the threat to endangered plant and animal species.

# 7 Soil

...l importance to countless living organisms and its nature
... many processes interacting over a long period. Since a
ch... just one factor may have far reaching effects, no two areas
have identical soils. Some important interactions are shown in Fig. 7.1.

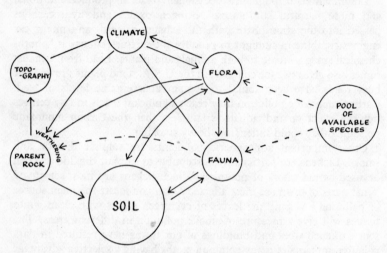

**Fig. 7.1**  Interactions affecting the soil

Many higher plants are characteristic of certain kinds of soil. The
ecologist asks 'Why?' and tackles this problem from three angles. He
studies the physiological capabilities of the plant at various stages in
its development, the physical and chemical structure of the soil, and
the influence of other organisms on these first two factors. Endogean
organisms, which live in the soil and rarely appear at the surface, are
particularly important. Bacteria, protozoa, and the slightly larger
mites, springtails and fungi all fall into this group.

Soil formation begins with the **weathering** of rock to give a framework of large and small mineral particles, with water and air in the
spaces. Ice, water, wind or gravity often transport this material so
that it no longer corresponds to the underlying bedrock. In many
places biologists will be unable to examine the parent-materials of the
soil, so geological maps should be consulted.

Weathering includes both physical and chemical processes.

PHYSICAL WEATHERING is the mechanical breakdown of rock into small particles by such agencies as frost (expansion of freezing water), and abrasion by water, wind or ice-borne particles.

CHEMICAL WEATHERING produces materials unlike the original rock; most important is the action of water, particularly when it contains dissolved carbon dioxide, causing hydrolysis, hydration or solution of minerals. Other important reactions are oxidation (as in conversion of ferrous iron to ferric) and the action of carbon dioxide producing carbonates. Biological processes may assist physical weathering as when roots penetrate and enlarge small cracks, chemical weathering

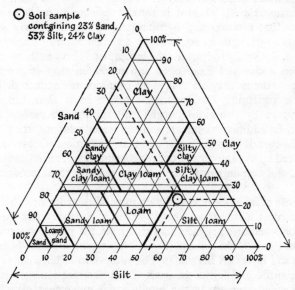

**Fig. 7.2** Textural designations based on mechanical analysis. (After Comber (1960))

as in respiratory production of carbon dioxide, or both, where burrows and decayed roots provide channels facilitating the weathering actions of water and atmospheric gases.

Weathering results in an assorted mixture of particles; the proportions of sand, silt and clay determine the textural designation of the soil.

Texture is important, because mechanical composition determines the volume and dimensions of the **pore-space**. These in turn determine the relative proportions of water and air, and the rates of water movement by drainage or capillary action. A sandy soil has a more open system of pore-spaces providing rapid drainage and better aeration, whilst a clay soil, though it may have a greater total pore-space volume, has a tighter system which reduces drainage and encourages capillary uptake and retention of water. Clay particles also hold water by colloidal imbibition so that soils with a high proportion of clay tend to be wet and heavy, whilst sandy soils are dry and light. Because water has a high specific heat, wet soils are affected less by climatic temperature change and warm up or cool down more slowly than dry soils.

Soil formation is completed by the production of **humus**, through the action of bacteria or fungi on dead vegetation. Where aeration is very good, decay is rapid and little humus is formed. With poor aeration, decay organisms may be so inhibited that the plant residues accumulate as peat. Humus is formed in conditions of reasonable moisture and aeration and is important because it returns valuable nutrients to the soil. In addition, its colloidal nature improves water retention in a sandy soil, whilst the incompletely broken down material will keep a clay soil more open and workable, thus improving the drainage. Humus may also act as an insulator and reduce daily temperature variation. The organisms active in it alter the soil–air composition by using up oxygen and raising the carbon dioxide content. In addition to climate and soil water relations, the nature of humus depends partly on the kind of vegetation. Decay rates are influenced by the amounts of decay resistant waxes, resins, tannins and lignins, and certain plants may contain more of these products than others.

Because of its mode of origin with mineral particles from below and organic constituents from above, soil is horizontally stratified. The layers or **horizons** progress upwards from the parent material or C horizon, through the subsoil (B horizon) to the organically enriched surface soil layers (A horizons). The succession of horizons is known as the **profile** and forms the basis of soil classification.

In the development of the profile water movement plays a great part, either by **leaching** out soluble substances from the A horizons and depositing them lower down during drainage, or less frequently by **flushing**, in which substances are washed upwards and deposited nearer the surface (near springs, in fenland or water-meadows). Leaching is most pronounced when rainfall considerably exceeds surface evaporation. The leaching of bases causes the soil to become more acid and this, if not neutralised by chalk or limestone, increases the losses by promoting the solvent action of rainwater. Intensive grazing removes transpiring tissues and can so alter the balance between rainfall and evaporation that leaching and acidity increase particularly over soil rich in siliceous sand. Under these conditions there are fewer bacteria and less nitrification, decay being effected mainly by fungi. Valuable nutrients may be removed entirely or deposited inaccessibly in the B horizons; here sesquioxides of iron, yellow to red in colour, are conspicuous and may under certain conditions combine with humus products to form a **hard pan** which restricts drainage and root growth. Above a hard pan the soil often becomes waterlogged so that only bog plants can grow.

Heavily leached **podsol** soils of this type with an acid humus, or **mor,** are found particularly on heathland, under pine forest, or in habitats where extreme conditions of temperature and moisture

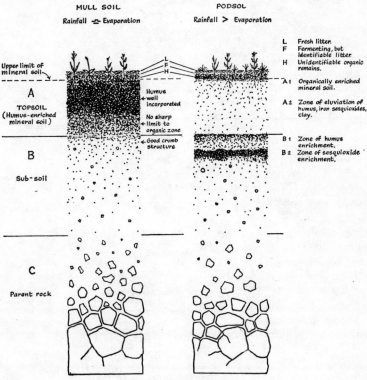

**Fig. 7.3**    The soil profile

inhibit bacterial activity. In most of the south and east of Britain, the milder climate results in a closer balance between rainfall and evaporation. These conditions favour a moist well-aerated soil in which bacterial decay forms a mild humus or **mull**. There is much less leaching and the soil is less acid.

The bacteria which break down dead vegetation are favoured by alkaline or near neutral conditions; this is one reason why liming is so important a procedure in agriculture (the other reason is that lime flocculates colloidal particles and improves aeration and drainage by promoting 'soil crumb' formation in clay soils). In acid conditions fungi are the principal agents of decay. Simple culture methods can be used to investigate whether fungi or bacteria are more abundant in a given soil.

Certain bacteria and fungi may form special symbiotic relationships

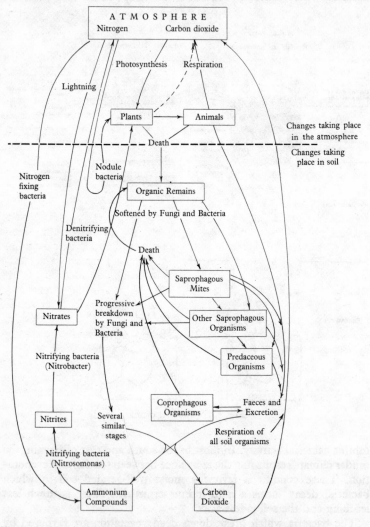

**Fig. 7.4** The part played by soil organisms, particularly the Mesofauna, in the nitrogen and carbon cycles. Organisms feeding on bacteria and fungi have been omitted

with the roots of higher plants. Root nodules containing nitrogen-fixing bacteria are commonly found in the Papilionaceae, and occasionally in many other plant families. Within one species there may be considerable variation in the number of nodules according to habitat. The same kind of variation is seen in **mycorrhiza** which are fungal associations with roots. What factors favour the development of these associations ? Are they the same for both fungi and bacteria ? The answers can only be found by systematic investigation and experiment. Both associations increase plant nutrient uptake, the invading organisms obtaining energy foods from their hosts.

Some mycorrhizal fungi are **endotrophic**, actually penetrating the cells of the host, whilst others are **ectotrophic** and never enter the root cells. Mycorrhiza are more common in mor soils than mull. These associations are very similar to those found in the case of fungi causing root rot, and in some cases the organism may change from symbiont to parasite under different ecological conditions and with different hosts. Except in autumn when many produce fruiting structures such as toadstools, puffballs, and truffles, all these fungi are difficult to identify. Predaceous fungi capable of trapping small animals like nematode worms are also occasionally present.

Diatoms and blue–green algae may be found in the surface layers where light is available for photosynthesis. Wherever there is organic material, there is an immense population of animals ranging from microscopic protozoa and rotifers living in the water films surrounding soil particles, through the mesofauna (roughly 0.2–2 mm long) to the macrofauna of beetle larvae, centipedes and millipedes, slugs and snails, earthworms, woodlice and so on. The variety is so great that exact identification requires specialized knowledge; however, identification to the general group is not difficult. Many of these organisms assist bacterial or fungal decay and are thus concerned in the circulation of nitrogen and carbon in nature (Fig. 7.4).

Some of these animals are 'burrowers' whose ability to excavate makes them independent of existing soil cavities. They show a wide range of adaptations for digging but convergent evolution has taken place between such dissimilar groups as mammals and insects where there is a remarkable resemblance between the fore-limbs of the mole and the mole-cricket (Fig. 7.5). As might be expected, an easily worked soil contains more animals than a heavy soil because burrowing forms can make their way through it more readily, and it is warmer and better aerated. The 'non-burrowers' include both the water fauna and small animals living in air spaces. Many of the latter may be found in other dark humid places like leaf litter and rotting logs. However, some of the litter inhabitants such as woodlice are too big to penetrate the soil proper and are restricted to woodland communities and sheltered places under stones.

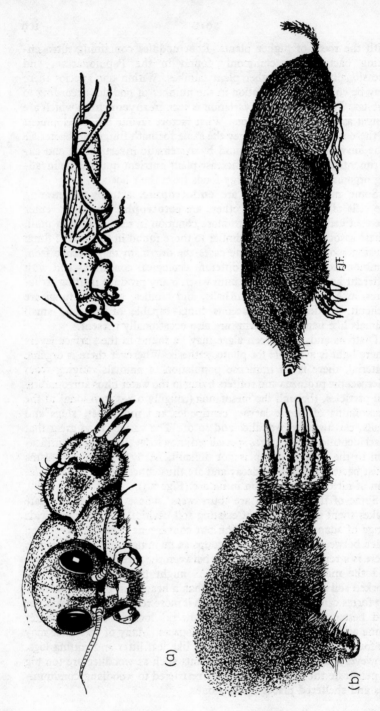

**Fig. 7·5** Convergent evolution between the fore-limbs of (a) the mole-cricket, and (b) the mole.

Some aspects of soil are very uniform, so that it is not surprising to find that animals from quite different groups have many features in common. For instance, reduced skin pigmentation and lack of eyes are related to the absence of light. If exposed to light, true soil inhabitants move away from it and in natural conditions this response returns them to the soil. Temporary inhabitants, however, are often attracted by light as when leaving the soil after hibernation.

Exchange of gases with the atmosphere takes place mainly by diffusion, and because this is slow carbon dioxide concentration tends to build up whilst the oxygen level remains lower than in the outside air. The resistance of animals to an increase of carbon dioxide tends to be related to the depth at which they are found in the soil; some larvae (*Melolontha*, *Agriotes*) which feed on plant roots may even be attracted by carbon dioxide. Hydrogen sulphide and ammonia, other products of metabolism and decay, also affect soil organisms. The effects of ammonia could be investigated using different strength ammonia solutions on trial plots; the widespread use of ammonium compounds as fertilizers gives added importance to this kind of study.

Most soil organisms, like earthworms, breathe through a moist skin and cannot resist desiccation, migrating vertically or horizontally if this danger threatens, as it may do in a sandy soil. Leaf-litter and topsoil contain animals which can, to varying extents, withstand drying and leave the soil from time to time. Seasonal differences in numbers and types of animals may well be related to changes in soil moisture content, as well as to the disappearance of insect larvae and pupae when they reach the adult stage. Animal frequency and variety are greatest where the air is permanently saturated. Where the soil is exceptionally moist or liable to flooding, animals may have difficulty in getting sufficient oxygen for their needs. These requirements may be reduced by the lower temperatures that generally accompany flooding, and in any case air bubbles trapped in soil cavities ensure some survivors. A few soil insects (some ants, beetles, collembolans) have non-wetting bristles on their bodies; these trap a thin layer of air which functions as a physical gill in making exchanges with the dissolved gases in the water. In contrast to flooding, desiccation raises particular dangers for the water fauna, and many protozoa, rotifers and nematodes respond by encystment. In some snails and rotifers the alternation between wet and dry conditions appears to be essential.

In the depths of moist soil, the temperature remains remarkably constant so that population changes are not markedly seasonal; at the surface the daily and seasonal variation is greater. Exchange of radiant heat is most rapid in dark coloured soils though actual temperature change is slower where the soil is wettest. Cooling due to evaporation is reduced where the surface is covered by vegetation or leaf-litter, or where soil particles are large enough to interrupt capillary pathways.

# Classification

| PROTOZOA | Rhizopoda | Amoebae | 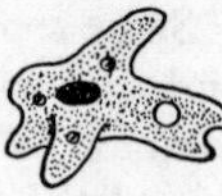 |
| --- | --- | --- | --- |
| | | Testaceous Rhizopods |  |
| | Mastigophora (Flagellates) | | 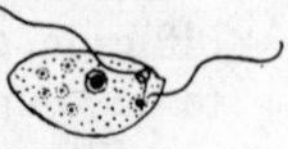 |
| | Ciliata | |  |
| ROTIFERA | | |  |
| NEMATODA | (roundworms) | |  |
| PLATYHELMINTHES | Turbellaria (flatworms) | | 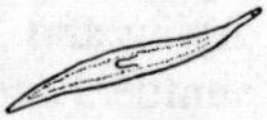 |
| ANNELIDA | Oligochaeta | Lumbricidae (earthworms) |  |
| | | Enchytraeidae (potworms) | 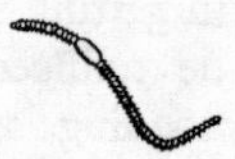 |
| ARTHROPODA | Arachnida | Oribateid mites | 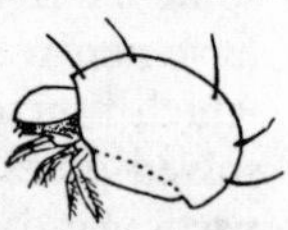 |
| | | Mesostigmatid mites | 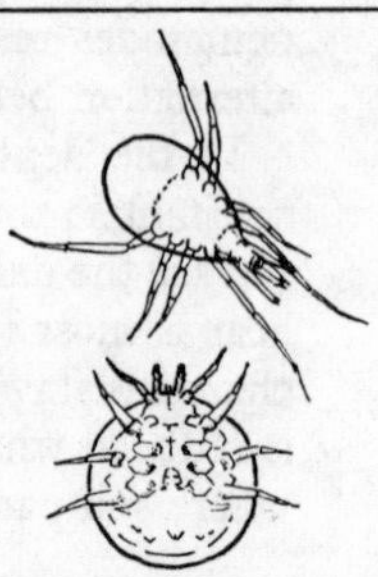 |

Fig. 7.6 Soil organisms. (Diagrams not all to same scale.)

| Diet and Feeding Habits | Where Found |
|---|---|
| Bacteria, larger micro-organisms, organic debris. | Water film<br>(Also in sewage filter beds) |
| As above | Water film, particularly in wet acid soils where mosses abound.<br>(Also in sewage filter beds) |
| Photosynthetic, saprophytic, or both (e.g. Euglena) Sometimes feeding on bacteria | Water film |
| Bacteria, small organic particles | Water film<br>(Also in sewage filter beds) |
| Mainly plant debris, but some feed on algae, protozoa, or nematodes. | Water film |
| Mainly: 1. Bacteria, fungi, algae, decaying plants.<br>2. Living plant roots (parasitic eelworms)<br>3. Other soil animals (predaceous round-worms) | Water film<br>Occasionally found trapped by predaceous fungi (cellophane film experiment—page 124) |
| *Rhynchodemus* feeds on small molluscs and annelids<br>Smaller turbellarians feed on microscopic animals of water film | Under logs, decaying vegetation, damp moss<br>Water film |
| Bacteria, fungi, decaying plant materials. Can digest cellulose; some digest chitin. | Burrowing; they strongly affect the soil flora. Rare in soils which lack calcium, are subject to drought, frost or waterlogging, or are acid. |
| Fungi, nematodes. (potworms may control early stages of eelworms) | Soil cracks, rarely burrowing. Potworms are small, white, and may be mis-identified as juvenile earthworm |
| Bacteria, fungi, decomposing plant remains. | Pore spaces |
| Predaceous on earthworms, potworms, nematodes, other mites. Some are ectoparasites of vertebrates, and many are parasitic on insects. | Pore spaces |

| ARTHROPODA<br>*continued* | Arachnida<br>*continued* | Pseudo-scorpions | |
|---|---|---|---|
| | | Araneida (spiders) | |
| | Crustacea | Isopoda (woodlice) | |
| | Insecta | Collembola<br>(springtails) | |
| | | Orthoptera | |
| | | Coleoptera (beetles) | |
| | | Diptera (flies) | |
| | | Hymenoptera | |
| | | Lepidoptera (moths and butterflies) | |
| | Myriapoda | Diplopoda<br>(millipedes) | |
| | | Chilopoda<br>(centipedes) | |
| MOLLUSCA | Gastropoda | Pulmonata (slugs and snails) | |

**Fig. 7.6** contd.   Soil organisms. (Diagrams not all to same scale.)

| Diet and Feeding Habits | Where Found |
|---|---|
| Predaceous on mites and collembolans | Pore spaces |
| Preying on insects and other small animals. | Leaf litter, soil crevices, under stones, occasionally digging burrows. |
| Scavenging on dead plant or animal matter. | Leaf litter, soil crevices, and under stones |
| Bacteria, fungi, decomposing plant remains | Pore spaces |
| *Gryllotalpa* (mole cricket) feeds on plant roots, tubers, other insects. | Active burrower |
| Many larvae (as shown) and adults with wide range of feeding habits, some attacking plants (wireworms, weevils), others scavenging (dor beetles), or predatory (ground, rove, and staphylinid beetles). | Active burrowers and channellers |
| Mostly larvae feeding on decaying or living plants (Tipulidae: leatherjackets) or parasitizing earthworms and molluscs. | Pore spaces |
| Mainly ants, preying on immature macrofauna. Some feed on plant material. | Active burrowers |
| Mainly non-feeding pupae, but a few larvae feed and live in the soil | Crevices and pore spaces |
| Most digest cellulose and eat decaying plants; some are pests of root crops. Are important for their part in mechanical breakdown of humus | Leaf litter, pore spaces and crevices. Some have larvae which may be confused with insects. |
| Preying on immature macrofauna | Leaf litter, crevices and pore spaces. |
| Feed at or above the soil surface on leaf litter, fungi, and sometimes on living plants. *Testacella* (a slug) is carnivorous, preying on earthworms. | Leaf litter in woodland. Elsewhere found in the larger crevices and pore spaces of the upper soil, or under stones. |

Soil animals are more susceptible to high temperatures than low and escape by vertical migration. Though they leave the surface in frosty weather, they may stay there if snow falls before the ground freezes. Many animals show temperature preferences which can be investigated by recording winter migrations or by comparing population levels at a particular depth on north and south slopes, or in shaded and unshaded patches.

Soil pH is less important to these animals than the factors already mentioned because there may be considerable variations within quite a small compass, related to the activities of roots or of other organisms. Most animals favour more or less neutral soils (pH 7.0), though in calcareous soils earthworms are able to neutralize carbonic acid by forming calcium carbonate in their calcareous glands. Thus though earthworms are more sensitive to pH than many other animals, calcium deficiency is even more limiting to them. Very acid soils such as the deeper layers of peat have no fauna, and in coastal areas salt spray or sea-water flooding may limit the number of species. Calcium compounds, needed to meet the skeletal requirements of woodlice, millipedes and snails, are less commonly a limiting factor.

If comparable soil samples are examined by standardized methods it soon becomes obvious that the distribution of soil organisms is distinctly patchy. To get even an approximation for population in an area the average of several samples must be taken. In many places, particularly in grassland, it will be found that the distribution of flowering plants above ground is also uneven. This leads to the question: 'Does one distribution reflect the other?' There is no reason why careful investigation should not give at least part of the answer. It is also necessary to discover what effect the soil animals have on each other. Which are vegetarians, and which carnivores and scavengers? By the time you have got some of the answers, it should be possible to piece together at least part of the food web and see the way to finding out more.

How do external factors affect the different animal groups? Much experimental work is waiting to be done. Many of the answers may be determined using experimental plots and treating them differently. Thus water content can easily be raised by regular use of a watering can, whilst on other plots lime, dung, weed-killers, insecticides and soot could be applied, and the soil fauna sampled at frequent intervals. Where possible there should be several plots of each kind and results and conclusions should be based on as many samples as possible because of the natural tendency to patchiness. The soil should also be sampled at different depths to discover what migrations, if any, are taking place. The more ambitious experimenter might try the effect of altering soil structure by mixing in sand or try to investigate how particular burrowing animals affect the soil and its inhabitants. What kind of food web is associated with an ants nest? How do burrow mammals affect the soil and the plants growing in it? Ecologists ha long recognized the changing flora associated with rabbit or badger communities and have tended to assume that it is due to an increase in nitrogen content as a result of excretion; it is, however, difficult to demonstrate any significant increase in nitrogen and the cause may well be some other mineral salt or even changes in aeration or drainage. Nettle growth is known to be associated with high phosphate availability. How do the soil animal and flowering plant populations of old mole-hills compare with adjacent populations?

Some of the most obvious effects of soil conditions are seen in the surface flora. Certain factors are particularly important; many plants are particularly sensitive to the presence or absence of calcium. Plants so characteristic of calcareous soils that they can be regarded as 'calcium indicators' are said to be **calcicolous**, whilst there are other **calcifuge** plants whose appearance indicates the absence of calcium.

Water content and pH are also of great importance. The remainder of this chapter is devoted to simple methods of investigating the more important soil factors.

## Investigating soil factors

### 1. *Soil profile*

Soil profile is best investigated by digging a vertical trench with a cleanly cut edge, and recording each horizon in turn down to the parent material. If this is not possible, auger or apple-corer type samplers may be used to remove a series of 15 cm long cylinders of soil which are then laid out in correct sequence and recorded.

When recording, a scaled sketch of the whole profile should be made, and a description of each layer given. This should include:

(*a*) Name of horizon, its depth and how clearly its boundaries are defined. Measurements should be taken from the surface of the *mineral* soil (top of the A horizon). Thus depth of L (litter), F (fermentation) and H (humus) layers is measured upwards, and other measurements downwards.

(*b*) Colour, and whether uniform or otherwise. Colour is usually due to humus (brown, grey, black), iron compounds (mainly red, orange or yellow) and calcium compounds (white).

(*c*) Texture (from Fig. 7.2 and section 3 below).

(*d*) Coarse skeleton: proportion and nature of stones present.

(*e*) Soil constitution: crumb structure as judged by handling when air-dry.

(*f*) Organic matter: its nature and distribution; whether structural remains of plants may be distinguished, and whether it is incorporated with the mineral particles of the A horizon. Raw humus often has a clearly defined lower boundary, usually contains plant remains, has no crumb structure, and has an acid reaction (pH less than 4.5). Mild humus (mull) has a diffuse lower boundary with no plant remains, a crumby structure and a less-acid to neutral reaction (pH 4.5–7).

(*g*) Root development: this is important for the clues given about moisture or aeration.

(*h*) Moisture: both moisture content (see section 5 below) and drainage, which may be described as excessive, free or poor.

## 2. *Sampling*

The ideal is to find a method which preserves the soil in its natural condition; the most usual methods are:

(*a*) Careful digging using a sharp knife or trowel.

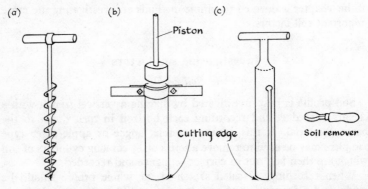

Fig. 7.7 (*a*) Soil auger. (*b* and *c*) Apple-corer type samplers

(*b*) Use of a soil auger. Because this is screwed into the soil, the sample obtained is of reduced value for measurements involving consideration of pore-space. The auger is screwed 15 cm into the soil and then removed by pulling directly out. The soil collected in the thread of the auger is removed and placed in a labelled polythene bag. The auger is then used to obtain another 15 cm sample from the same hole, and so on. Deep samples can be obtained more readily by this method than by the others.

(*c*) Apple-corer type of sampler. This consists basically of a cylinder with sharpened end which is driven into the soil and then rotated with a screwing motion to break the soil at the mouth of the cylinder and remove a sample. The sample is forced out of the sampler by some sort of piston arrangement and is collected as a cylinder of soil retaining much of its natural structure. A soft soil may be compressed by the piston, but allowance can be made for this by comparing the length of the sample with the depth of the hole. With sticky soils it is difficult to obtain samples more than 15 cm long owing to sticking in the sampler. All these methods may be used to obtain samples horizontally from the vertical face of a profile pit, but in this case the first 15 cm sample should be rejected to allow for moisture loss and fauna migration.

### 3. *Mechanical analysis*

The relative proportions of sand, silt and clay are determined by sieving to remove the sand, and then separating the silt from the clay by differential settlement in water. A weighed sample of oven-dried soil is passed through a 2 mm sieve to remove stones, then through a very fine sieve to separate the sand fraction which is removed and weighed. The particles which have passed the fine sieve are collected in a large, weighed, evaporating dish. After reweighing, distilled water is added to within 1 cm of the top and the liquid thoroughly stirred. It should then be allowed to stand for several hours before decanting off the cloudy supernatant liquid. The process is repeated until the decanted liquid is quite clear – usually three stirrings are enough. The silt remaining in the evaporating dish is then evaporated to dryness and weighed; the clay fraction which has been decanted is determined by subtracting this value from the previous weighing. Quite reliable values are obtained in this way and can be used with Fig. 7.2 to find the textural designation.

### 4. *Organic matter content*

When strongly heated in air, organic matter is oxidized mainly to carbon dioxide and water. Thus the loss in weight when oven-dried soil is heated to red-heat represents organic matter and is expressed as percentage by weight. The normal precautions associated with gravimetric determinations need to be taken, the sequence being as follows: Heat a crucible and lid strongly to drive off moisture; cool in a desiccator and weigh. Add oven-dried* soil (about one-third full) and reweigh. Heat to red-heat for at least an hour. A crucible lid is essential to avoid 'spitting' during heating. Cool in a desiccator before weighing. Heat again for 10 min and reweigh; repeat this until successive weighings are identical. The calculation is straightforward.

Weight of crucible $= x$ g
Weight of crucible + soil before heating $= y$ g
Weight of crucible + soil after heating $= z$ g

* Dried at 100 – 105°C

Weight of original soil                    $= y - x$ g
Weight of organic matter                   $= y - z$ g

$$\text{Percentage organic matter} = \frac{y-z}{y-x} \times 100$$

## 5. *Soil water*

Water content of a soil varies between wide limits. This means that a direct determination of the moisture present in a soil sample may have little bearing on the distribution of flowering plants which are present for long periods ranging from months to years, though it may well throw light on the migrations of the soil fauna. For many purposes a determination of **field capacity** may be more meaningful as it does not depend on the weather just before sampling.

(*a*) TOTAL WATER PRESENT is determined by weighing and then heating to constant weight at just over 100°C in an oven. At this temperature all the uncombined water is driven off, but organic matter is not destroyed. Some of the water driven off by this method is hygroscopic water which is not available to plants at normal temperatures. The moisture, as also in the following measurements, is expressed as a percentage of the total sample weight.

(*b*) AVAILABLE WATER is determined in a similar way, but by *air* drying a weighed sample to constant weight at *room temperature*.

(*c*) FIELD CAPACITY is the maximum amount of water which a freely drained soil can hold. To determine this, the soil should be in its natural condition – neither broken nor compressed. A simple method is as follows. Prepare a simple corer type sampler by cutting off the roll-edged end of a tall slim can (e.g. soft-drink or beer can) with tin snips or a rolling type can-opener used sideways. Weigh the can, which now has a sharp cutting edge. Perforate the other end and drive the can into the ground until the perforated end is flush with the surface; the soil will not be measurably compressed. Carefully dig the can out of the ground, wipe its outer surface clean, and weigh. Lower the can into water perforated end first, leave it until the soil is thoroughly soaked, and then remove and allow to drain freely. When water ceases to drip, dry the outside of the can carefully with blotting paper and weigh. Any increase in weight is due to water uptake, and field capacity may be determined either by oven-drying and weighing, or by combining the result of this experiment with that from experiment (*a*) above.

*Example:* If the increase in weight after soaking and draining is 20% of the original soil weight, 100 g of soil will take up 20 g water.

If the result from (*a*) was 40%, the 100 g sample of soil will contain 40 g of water before soaking.

That is, after soaking the sample weighs 120 g and contains 60 g of water – a field capacity of 50%.

(d) MAXIMUM WATER CAPACITY is the maximum amount of water the soil can hold if drainage is prevented and all air displaced; it is equivalent by volume to **pore space**. This may be determined as follows: Collect a soil sample as above, cutting the soil off level with the open end of the can; weigh it. Take a wide measuring cylinder or gas-jar and fill it with water to a level which will just cover the can; mark this level $(x)$. Lower the can and sample into the water and carefully excavate the soil to displace the air; mark the new level of the water $(y)$. After washing out the can and cylinder, refill the cylinder to the original level $(x)$, and, after sealing the perforations with adhesive tape or paraffin wax, fill the can to the brim with water and lower it into the cylinder. Mark the new level $(z)$. Finally determine the volume of air displaced by using a measuring cylinder to determine how much water must be added to raise the level from mark $(y)$ to mark $(z)$.

Assuming 1 cm$^3$ of water weighs 1 g, $(z-y)$ can be expressed as an increase in weight and the maximum water capacity calculated in the same way as field capacity in $(c)$ above. Alternatively, using the results from $(a)$ above, the original water present can be estimated in terms of volume and added to the volume of air displaced $(z-y)$ to give pore space in terms of percentage by volume.

### 6. Soil atmosphere

Methods for accurate analysis of the soil atmosphere are beyond the scope of this book. However, it is not difficult to make rough estimates of oxygen and carbon dioxide percentages using absorbents such as strong potassium hydroxide solution for carbon dioxide and strongly alkaline pyrogallol solution for both carbon dioxide and oxygen. The air sample should be collected by burying a pair of tubes in the soil and leaving them long enough to reach equilibrium with the soil atmosphere (about a month). Because of the danger of breakage, thick-walled transparent plastic tube is better than glass. One end of each tube has a rubber bung fitted with capillary tube and a simple stopper, whilst at the other is a bung fitted with a rubber tubing outlet which is closed with a screw clip before removing the sample from the soil. Because the sampling tube is flexible, it may be deformed to force air out of the capillary outlet and into a gas burette for analysis. An account of this method, using a Barcroft–Haldane apparatus, is given by M. G. Brown and W. H. Dowdeswell (1956, 'Apparatus for the Ecological Study of Soil and Mud', *School Science Review*, No. 134).

A single sample is sufficient for capillary-tube gas analysis as outlined in either Text III (*The Maintenance of Life*) of the O Level Nuffield Biology Course, or the laboratory guide *Maintenance of the Organism* associated with the Nuffield Advanced Biology Course. Here, a column of gas, sealed by water at both ends in a capillary tube, is measured (length A) and then exposed to potassium hydroxide. After recording the new length (B), the column is exposed to alkaline pyrogallol, resulting in a still shorter length (C). These three lengths are used to calculate the percentage of carbon dioxide $\left(\dfrac{(A-B)}{A} \times 100\right)$, and percentage of oxygen $\left(\dfrac{(B-C)}{A} \times 100\right)$. Care must be taken to ensure that each exposure has lasted long enough to ensure full gas absorption, and that each measurement is made with the tube at the same temperature.

## 7. *Soil reaction*

Soil acidity or alkalinity is measured on the 'pH scale' which is a method of expressing hydrogen ion concentration. On this scale, 7 is neutral, values below 7 are acid and values above 7 are alkaline. The amount by which the value differs from 7 gives a measure of the strength; thus, pH 8 is mildly alkaline, pH 6 mildly acid, and pH 3 strongly acid. There are various methods of measuring soil pH, most of them based on colour changes in indicator solutions. These are:

(i) Place about 2 g of soil on a crucible lid, soak it with BDH Universal Indicator, wait 1–2 min and then tilt the lid so that indicator drains out of the soil. Compare the colour with the chart supplied with the indicator.

(ii) Proceed as in (i) but soak the soil with distilled water. Drain, then test the water with Universal Indicator Test Papers – then get a more accurate value using a paper with a more limited range.

(iii) BDH Soil Test Outfit. This uses tubes in which a small quantity of soil is mixed with barium sulphate, water and indicator and shaken. The barium sulphate ensures flocculation and precipitation of colloidal clay but may itself give rise to inaccuracy due to acidity if not of guaranteed quality. The resulting coloured solution is compared with a colour chart.

(iv) BDH Capillator Sets are available for various indicators covering more limited pH ranges and giving more accurate values. Here mixed soil solution and indicator is drawn into a capillary and compared with standard indicator solutions.

(v) Use of a Lovibond comparator to compare indicator colours with glass standards (see also Chapter 6, p. 99).

(vi) pH Meters provide direct and accurate pH readings, but are rather expensive and need some preparation before use.

## 8. *Carbonate content*

Add a little water to about 5 g of soil in a boiling tube, and shake to remove air bubbles. Add 10 cm³ of dilute hydrochloric acid. If carbonate is present (mostly calcium carbonate) there will be effervescence – marked if $CaCO_3$ exceeds 1%; very slight if less than 0.5% is present.

A more accurate quantitative estimate can be made by treating a weighed soil sample with standard hydrochloric acid; when there is no further reaction the remaining acid may be estimated by titrating with standard sodium hydroxide.

## 9. *Soil organisms*

(*a*) MACROFAUNA. Larger arthropods, earthworms, slugs and snails may be collected by hand, with or without the aid of a sieve (see also use of repellents, Chapter 3).

(*b*) MESOFAUNA. Smaller organisms may be extracted using Tullgren or Baermann funnels or flotation (see Chapter 3, p. 30*f*).

(*c*) MICROFAUNA. A useful qualitative method is to flood a 2% agar plate in a Petri dish with hay infusion to a depth of 1 cm. Spread small samples of soil thinly over part of the agar surface; cover and leave undisturbed. Protozoa emerge and explore the agar, where they may be observed under the microscope.

For quantitative determinations make a dilute suspension of soil in water (about 4 g/litre), and filter through glass wool or coarse filter paper to remove most particles. Spread about 1 cm³ of the filtrate over one end of a microscope slide and dry over moderate heat. The dry film may now be stained (e.g. with carbol–fuchsin) and examined under a high power microscope for protozoa, smaller metazoa and bacteria.

(*d*) MICROFLORA. Saprophytic fungi and bacteria may be identified and counted by making dilution and crumb plates. A weighed soil sample is diluted with sterile water to 1/10 000 for fungi, or 1/250 000 for bacteria. Do this by shaking the soil with a small volume of water (say 10 cm³); take 1 cm³ of this solution, dilute ten times and shake. In this way, if the original water contained 1 cm³ of soil, the required strength of 1/10 000 is reached with only three subsequent dilutions. With the recommended strengths, each colony developing in culture is likely to have arisen from a single spore or bacterium. An agar plate is made by pouring sterile molten medium at 45°C into a sterile Petri dish containing 1 cm³ of the final dilution. Alternatively a small soil sample (one or two crumbs – about 15 mg) may be dispersed directly in the agar medium. Useful media include Malt Agar, Potato Dextrose Agar, Oatmeal Agar, Czapek-Dox Agar, Soil Extract or Dung Extract

Agar – the last being particularly suitable for isolating some of the slower growing organisms. Recipes for these media are given in the Appendix. Fungi require sugars and an acid (pH 5) medium whilst bacteria usually require products of protein hydrolysis and a neutral to alkaline medium. The plates should be incubated in warm moist conditions; colonies should begin to develop within the week. Many different media are available in powder or tablet form from Oxoid Ltd., who also supply a very useful handbook and buffer solutions for adjusting the pH.

(e) MICRO-SUCCESSION. If a simple organic substrate is buried, a succession of organisms will be involved in its breakdown, and the pattern of the succession will vary according to conditions. A suitable material is plain uncoated cellophane (available from D. J. Parry Ltd., Avonmore Road, London, W14 8UE) which has been boiled in water to remove plasticizers and then washed in distilled water. Each of a number of small pieces of prepared cellophane is placed in the centre of a clean coverslip. The slips are buried in flower pots by making vertical slits in the sieved soil. Four such slits may be arranged radially in one pot and cover-slips may be buried at different depths. Mark positions and depths on the outside of the pot. Arrange the slips so that each piece of cellophane faces clockwise; then the cover-slip may be removed without destroying the catch (see Fig. 7.8). The soil must be kept moist and the cover-slips excavated after varying periods (e.g. 2, 4, 6 weeks). Soil is scraped from the non-prepared surface and dislodged from the other side by tapping. The cover-slip may be fixed and stained, then washed and mounted on a slide for examination under

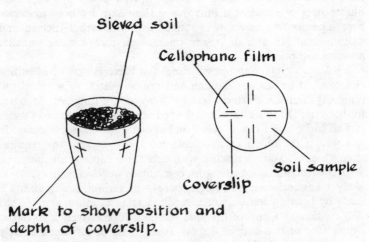

Fig. 7.8   Buried cellophane film. (*left*) marked pot; (*right*) top view to show coverslips set radially, with cellophane film always on clockwise face

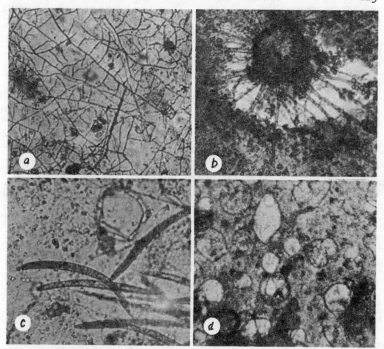

**Fig. 7.9** Organisms found after two (*a* and *b*) and four (*c* and *d*) weeks on buried cellophane film. (*a*) fungal mycelium is established; (*b*) an unusual fungal growth; (*c*) colonization by nematodes; (*d*) beginnings of a population explosion – nematode eggs and embryos are visible. (*d*, is at a higher magnification than *a* to *c*)

the microscope (e.g. stain and fix in lactophenol blue or 2% nigrosine in picro-lactophenol, wash and mount in lactophenol. Instructions for the preparation of the nigrosine stain are given in the Appendix). This technique may be used not only to compare the processes of decomposition in soils from different sources, but also to investigate the effects of herbicides or insecticides at different concentrations on the rates of decomposition and the organisms concerned. Further information about the technique, with many photographs and descriptions of organisms found at different stages in the succession, may be found in Tribe, H. T. (1967), 'Biological Decomposition in Soil', in *School Science Review*, No. 167, pp. 95–112. Organisms found in this way include fungi and bacteria as well as animals feeding on them or their products (arthropods, nematodes, etc.). As might be expected, soil samples from the F (fermentation) layer are most productive.

# 8 Woodland

Woodland affords excellent opportunities for observing the principles and practising most of the techniques outlined in earlier chapters. The most important habitat factor is light, the whole character of the community being dominated by shade-producing trees. These also act as windbreaks, and moderate other habitat factors so that the woodland atmosphere is more humid and temperature variations are less extreme than in an open habitat. Similar features are shown by readily available habitats like hedgerows and even shrubberies.

## Light

Woods dominated by a single species are more common than mixed woods. Oak, ash, beech, birch and pine are the main British types of woodland. The dominant trees of mixed woods are commonly oak and ash in the south, and birch and pine in the north. The degree and duration of shading varies greatly between the different tree species – thus heavy shading beginning early in the year, as in beech-woods, results in a very limited subordinate flora, whilst the later developing and more open canopy of oaks is associated with a rich assortment of shrubs and herbs.

Woodland plants require less light than the dominant trees which are **sun plants** and do best in high light intensities. **Shade plants** usually have a shorter **compensation period**, during which photosynthesis replaces energy food reserves used up by respiration in the dark. The **compensation point** is the light intensity at which losses from respiration are balanced by gains from photosynthesis and this may be very low for some of the mosses found in deep shade. A simple method for determining compensation periods of different plant tissues is given by W. M. M. Baron (*Organization in Plants*, 2nd edn., 1967, London, Edward Arnold).

Whilst light intensity is of great importance, the quality of the light must also be borne in mind. The green appearance of plants indicates that green is the least absorbed of all colours of the spectrum and is therefore least useful for photosynthesis. Under a closed canopy, green light predominates – as many photographers ruefully discover in their first attempts at photographing woodland in colour.

Light affects many plants by controlling 'sleep' movements which open and close flowers at certain times of the day. In the case of wood sorrel, *Oxalis acetosella*, it is the trifoliate leaves which fold at night-time or in direct sunlight when special cells at the base of each leaflet become less turgid; this is said to reduce the transpiration rate.

Measurement of light at any one spot is difficult because the illuminated regions change during the course of the day. Flecks of sunlight on the ground and herbage indicate how much white light penetrates the canopy. Such sunflecks are often of great importance for photosynthesis. Wind, too, by distorting the canopy, can affect light availability.

## Succession

Deciduous woodlands, losing the canopy in winter as a result of leaf fall, are characterized by a seasonal succession (see Chapter 2, p. 14) related to changes in temperature, rainfall, light intensity and day length. Many woodland plants flower, fruit, undergo maximum

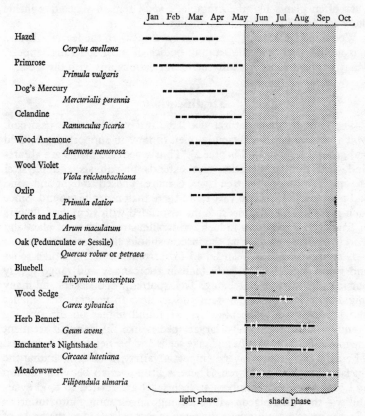

**Fig. 8.1** The flowering periods of woodland flowers in relation to light and shade

growth and accumulate food reserves for next year's flowering before
the foliage of the dominant tree limits available light; these plants are
least successful under 'early' trees, like the beech. Flowering in many
plants is affected by day length and the amount of incident light,
requirements differing so that the flowers of the ground flora succeed
one another through the year (Fig. 8.1). Relatively few plants are in
flower when the canopy is fully developed.

There is a difference in the duration of the leafy condition as well
as of time of flowering; both the wood anemone and dog's mercury
flower early, but the leaves of the anemone are short-lived, whilst
those of dog's mercury are semi-evergreen. How do these compare
with other herbs? Many trees and shrubs also flower early in the
spring, since the absence of leaves gives better chances of pollination.
The seasonal succession of seed and fruit production which follows is
reflected by the changing feeding habits of many birds and rodents.
Very often plants like the bramble, *Rubus*, remain vegetative inside
the wood but flower and fruit outside.

Woodland becomes lighter in the autumn when the leaves fall, and
it is at this time that the fruiting bodies of most soil fungi appear.
Many of these fungi have mycorrhizal relationships with the trees.

### Stratification

The most common natural woodland in Great Britain is oakwood,
which usually has a sufficiently open canopy to support shrub, field
and ground layers (see Chapter 2). Thus woodland plants are **stratified**, and this stratification also extends to animals and physical
factors. The canopy, or tree layer, is more exposed to light and wind
and temperatures tend to vary more there than near the ground. Since
each plant has its own insect fauna associated with stem, leaf, flowers
or fruit, it is a good idea to look for parallel stratification of plant and
insect. Animals preying on the insects should also be considered.

A stratification has been noted for certain birds. Birds like rooks
and wood pigeons nest fairly high in the canopy and range widely
outside the wood when feeding. Tits, spotted woodpeckers and many
migrant birds feed in the trees and shrubs. In general, the smaller
and lighter species like blue tits are found higher on the swaying
branches and twigs, whilst larger species like the great tit frequent
the more sheltered shrubs and take seeds and berries as well as insects.
The field and ground layers support a variety of birds including the
thrush, blackbird and wren. There is little overlap between the top
and bottom strata. Numbers at each level are related to food availability – the wood pigeon with a feeding range running into hundreds
of acres greatly outnumbers the wren which is limited to the herbs of
the field and ground layers.

Although many different birds feed in the wood, interspecific competition for food is reduced by specialized feeding behaviour and the great variety of beak adaptations for dealing with different kinds of food (Fig. 8.2). Even within a single species direct competition in the breeding season is avoided by division of the wood into separate territories, each owned by a single bird or mated pair. The behaviour of territorial birds is well worth investigating. Migration also regulates competition, for the steep drop in bird density in winter is related to the reduced availability of insects. Most woodland birds are insectivorous, although some will take to other foods in the winter.

**Fig. 8.2** Bird beaks. **Insectivores:** (a) woodpecker—stout, pointed for piercing bark; (b) blackbird and thrush—pointed for taking insects, worms, etc. from soil, less strong than (a); (c) tree creeper—fine, curved for exploring crevices. **Seed and fruit feeders:** (d) wood pigeon—fairly blunt, pecking for seeds; (e) cross bill—specialized for extracting conifer seeds. **Predators:** (f) kestrel—strong, hooked for tearing flesh

The most common kind of lowland oakwood on damp clay loam is 'coppice with standards' – oaks set fairly widely spaced with hazel as subdominant (Fig. 8.3). The hazel is coppiced every ten or twelve years, and in any one wood there may be three or four different ages of hazel coppice which can be compared for shading and other effects. The light available when the hazel is cut back leads to increased growth of the field layer, possibly inhibiting the ground layer of bryophytes. Where oak trees are felled a similar increase occurs, particularly of

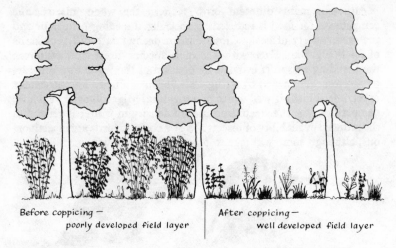

Before coppicing —
        poorly developed field layer

After coppicing —
        well developed field layer

**Fig. 8.3**  Coppice with standards, before and after cutting

biennial herbs which seldom restrict regeneration by young trees in
the way that scramblers or bramble do.

## Distribution

The distribution of herbs in the field layer is usually uneven; in
some cases this is because the plant is unable to grow in areas receiving
less than a certain percentage of full daylight which is known as the
plant's **extinction point**. Quite apart from this, vegetative spread
tends to produce continuous sheets of plants of the same species.
Stratification also plays a part, because it extends underground to the
roots (Fig. 8.4). Except for beech, which has roots in the surface soil,
most tree roots do not compete for water and mineral salts with the
roots of the field layer. In the field layer there may be competition
between different species with roots at the same level.

Distribution of trees depends on two aspects; dispersal and regeneration. The mechanism, pattern and extent of dispersal are
easily studied by direct observation. Regeneration can be studied both
by looking at natural seedlings, and by experimentally planting seeds
or seedlings in chosen places. For instance, ash seedlings seldom
become established beneath beech, since the dense canopy appears
too early in the year.

Animals generally occur in sites suited to their special feeding or
nesting needs. The abundant leaf litter found in deciduous woodland
forms the habitat of woodlice, centipedes, millipedes, collembolans
and spiders. What parts do these play in the annual cycle of events

which return nutrients to the soil? What changes take place in the population if birds are excluded from an area by wire netting?

Animal distribution on single plants can make a most interesting study. For instance, some of the organisms in the soil directly attack the roots of plants. Thus the fertilized female gall-wasp *Biorrhiza pallida*, a temporary soil inhabitant, crawls down the trunk of an oak tree from the 'oak-apple' where it hatched. It lays its eggs in the roots, where the larvae cause spherical galls. From these, wingless partheno-genetic females emerge in January to climb up and lay eggs in terminal

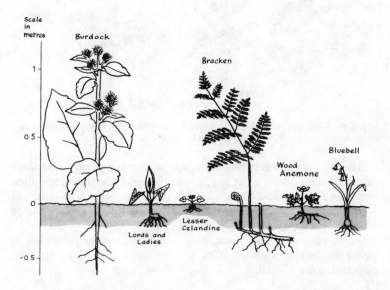

**Fig. 8.4**  Stratification in both roots and shoots of the field layer, reducing competition—particularly for water and mineral salt resources

buds; this results in oak apple galls and the cycle begins again. The oak tree is attacked by many other gall producing insects, causing galls in acorns and bark as well as in leaves and leaf buds (Fig. 8.5). Over fifty different gall forming insects are known to attack oaks; most of these are wasps, but there are also some moths, flies, aphids and mites which cause galls. Quite apart from gall formers, caterpillars of various moths (Buff-Tip, Mottled Umber, Winter-Moth, Oak-Roller) feed on the leaves and may pave the way for fungal attack. Tunnelling caterpillars attack the bark (Goat and Leopard Moths) or seedlings (Ghost-Swift Moth). Other insects living on the oak include aphids and scale insects, which do not appear to cause much harm, and beetles which attack both the living tree (weevils, bark beetles) and the rotting logs (stag beetle, furniture and death watch beetles).

It will be evident from this that the single oak tree has far reaching effects on organisms other than those affected directly by its shade. It is attacked by many fungi, including the 'beefsteak' fungus and many bracket fungi, and also provides a habitat for climbing plants and epiphytes like ivy and mosses.

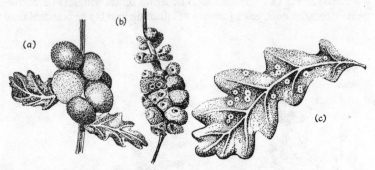

**Fig. 8.5** Some galls on oak trees. (*a* and *b*) *Biorrhiza:* oak apple and root galls. (*c*) Spangle galls formed by gall wasps

The oak produces hundreds of acorns in most years, but few of these produce young trees; what happens to the rest? Careful observation over one season should provide part of the answer. The study of seedling stages is particularly interesting, as competitive relationships are often shown very clearly.

Where squirrels are present, they may be the principal factor in destroying acorns. Paradoxically, they are often the most important dispersal agent as well.

## Scrub

Scrub precedes woodland in an ecological succession (p. 17) and resembles it in the enduring nature of its woody structure. It differs in its fine branching which makes it impenetrable by man, provides cover for mammals, such as rabbit and fox, and nesting sites for birds. Natural scrub is deciduous and exists as thicket, hedgerow, woodland edge, and the undergrowth in open-canopied woodland. Its dense canopy suppresses the development of herbs below with the result that there is usually bare soil. In Britain, less than a dozen common shrubs form scrub, aided by a few woody or semi-woody climbers such as honeysuckle, *Lonicera,* and old man's beard, *Clematis.* Which are the principal scrub formers in your area?

Most of the shrubs have flowers which attract insects, and produce fruits and seeds that are taken and dispersed by birds. They contrast strongly with woodland trees which are mostly wind-pollinated and wind-dispersed, and the shrub dispersal mechanism explains why a shrub undergrowth may develop in woodland where it can rarely flower and fruit successfully.

Hawthorn and elder are particularly successful scrub pioneers because of their resistance to attack by rabbits. They are also typical of hedgerows, hawthorn in particular being traditionally used to form a close hedge and windbreak. Such hedges provide shelter and influence neighbouring land to some distance. How do close hedges compare with more open hedges which include trees? How far does the shelter extend? Which type of hedge produces the most eddy currents, with resulting damage to crops in high wind?

Many of the animals found in scrub are there on a temporary basis – some briefly for feeding purposes, others over a longer period for nesting, or to overwinter. As already mentioned, scrub is likely to be supplanted by woodland unless the succession is prevented by such human interference as the cutting and laying of hedges – which is therefore a form of conservation.

## Productivity Measurement

Harvesting, drying and weighing the litter of bud-scales, flowers, fruits and leaves which accumulates in a specific area of woodland floor during the year can give a useful indication of primary productivity. It is usual to use a set of deep bags suspended from metal hoops or frames of known area; the bags should be of materials such as nylon mesh which drains freely, reducing moisture and decomposition. Collecting from quadrats marked out on the soil may yield misleading results because litter tends to drift in the wind and some kinds of moist litter decompose more rapidly than others. Bags should be emptied and contents sorted at least monthly. Droppings from caterpillars and other leaf-eating insects can provide a measure of their activities when compared with the droppings produced by the same insects reared in the laboratory. Woodland productivity and energy-flow are discussed on p. 232.

## Soil

The character of oakwood is by no means constant. Of the two species of oak common in Britain *Quercus robur* is more common in the south-east, whilst *Q. petraea* is found in the west and north. Descriptions so far have been limited to oakwood on a moist loam, but on other kinds of soil the subordinate vegetation may be very different. Thus, on a sandy and more acid soil there are fewer shrubs and the field layer often contains bracken. Many of the herbs common on loam are replaced by acid favouring plants like the foxglove *Digitalis purpurea*. If the soil is *very* acid, the oaks are adversely affected and the tree layer may include a high proportion of birch with few shrubs except ling and other heath plants, together with heathland grasses like *Festuca ovina* and *Agrostis tenuis*. In very wet conditions likely to become waterlogged in winter, alder is commonly present and may even replace the oak as dominant in places. Here sallows (*Salix atrocinerea, S. aurita*) occur in the shrub layer, and sedges, rushes, meadowsweet and creeping buttercup in the field layer.

Similar considerations apply to the dominant trees of other kinds of woodland. Beechwoods are most successful on the chalk of the Kent Weald and the Chilterns, or the oolitic limestone of the Cotswolds. Such woods contain few, if any, other kinds of tree, no shrubs except evergreens able to survive on the light available in early spring, and a sparse field layer usually dominated by dog's mercury and sanicle *Sanicula europaea*. As might be expected, the dense canopy provides a deep leaf litter in autumn. Beech leaves are more resistant to decay than those of other trees; half the leaves may still be recognizable a year after falling, compared with none at all for wych elm. On different soils, the beech may be less successful – thus on deep loams other trees are present and even frequent. The beech only produces good crops of seeds at irregular intervals and this may have some bearing on why ash is dominant in similar habitats in the north and west.

Birch and pine dominate woods in more exposed places like heaths and moorland, particularly in the north.

Trees fail to regenerate and may sometimes be destroyed as the result of grazing animals' activities or because of extreme exposure. The grasslands, heaths and moorlands which result are considered in the next two chapters.

# 9 Grassland

Whilst the hedgerow exhibits many features of woodland in miniature, the roadside verge shows many of the features of grassland and is equally worthy of study. It is, like most grassland, an example of a **deflected climax** vegetation maintained by biotic interference in the form of grazing, seasonal mowing, burning or weed spraying. In many ways, the verge is better for study than the nearby meadow which has probably been reseeded and contains a very limited number of species, few of which may be natural to the locality. Many of the suggestions in this chapter can apply equally to the roadside verge and to the mature grasslands of chalk downs, hill pasture and common lands generally.

Calcareous grasslands are richest in species. The diversity of the flora is largely due to grazing by sheep or rabbits, which reduces competition for light; there is a close association between the height of the herbage and the number of species present. This can easily be investigated either by experimentally excluding the grazing animals, or by comparison with an unmown and ungrazed road verge or hayfield. In general, the soil on chalk downs is shallow and the high lime content confers alkalinity which promotes bacterial action to produce mild humus. The soil is well aerated and porous so that, although the activities of burrowing animals stir up the subsoil and slow down the process, considerable leaching may occur and cause acid conditions at the surface. On moderately steep slopes, surface erosion takes place rather than leaching, so that lime content and alkalinity in the remaining soil are little changed. The floral differences between level soils and slopes lend themselves readily to investigation.

Availability of many plant nutrients is related to the lime status and pH of the soil. Thus lime increases molybdenum availability and this favours root nodule formation in pea-family plants. Manganese, however, can only be absorbed in acid conditions so that plants may show deficiency symptoms except where surface leaching has taken place. Many plants are deep rooted; these probably get one set of nutrients from the acid surface soil and another from the alkaline sub-soil. Root production is said to be stimulated by calcium ions; on chalk soils this is coupled with a low mechanical resistance to soil penetration. Root systems of plant species found on both chalky and non-chalky soils can be compared by carefully excavating and floating out the roots on glass plates; root-hairs should be excluded when measuring the total root length.

The most typical plants of grassland are perennials in which the growing point is protected from grazing animals by being close to soil level or underground, as in the case of bulbs, corms and rhizomes. In many plants the rosette habit, in which the leaves arise close together at ground level, both protects the plant against grazing animals and helps to reduce water loss by transpiration. In contrast, 'mat plants' achieve much the same result by spreading along the soil surface, sometimes putting down adventitious roots. On rosette plants,

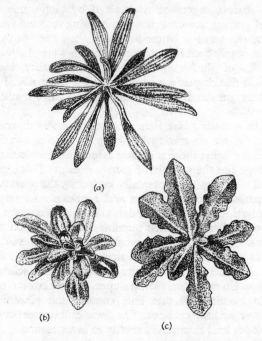

**Fig. 9.1** Rosette plants. (*a*) Ribwort plantain (*Plantago lanceolatus*). (*b*) Common daisy (*Bellis perennis*). (*c*) Cat's ear (*Hypochaeris radicata*)

lateral buds may also give rise to a kind of mat, as is sometimes seen with the common daisy on a neglected garden lawn.

Where chalk slopes are steep, the vegetation is usually open and the soil surface is incompletely protected by vegetation. In these dry conditions most successful plants show *xeromorphic characters* (see Chapter 10) – structural adaptations which limit water loss. Thus the most favoured grasses are those whose leaves can respond to water shortage by rolling or folding. A number of root parasites of grasses such as yellow rattle (*Rhinanthus*) and eyebright (*Euphrasia*) are often present. This is probably linked with the need for water and

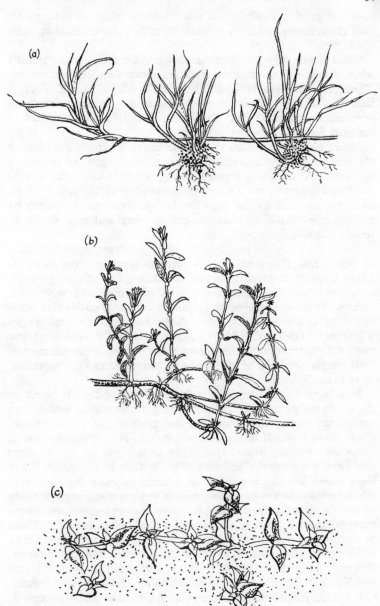

**Fig. 9.2** Mat plants. (a) Creeping bent (*Agrostis stolonifera*). (b) Mouse-ear chickweed (*Cerastium semidecandrum*). (c) Sea sandwort (*Honkenya peploides*—a perennial pioneer dune-builder, see Chapter 13)

dissolved nutrients rather than manufactured foods, as these plants have green leaves and some at least can grow independently in moist conditions.

Chalk downs support a very varied fauna; the chief grazing animals which determine the character of the vegetation are sheep and rabbits. There are numerous butterflies and moths whose caterpillar stages feed on the many kinds of plant. Snails and their predators are particularly common; thus glow-worm distribution reflects that of the snail *Helicella* on which it feeds.

Severe grazing pressure can result in almost complete denuding of the habitat. When this pressure is relieved, the vegetation returns in the same succession as is found on newly exposed chalk soils. Mosses, the first colonizers, are succeeded by tufted pioneer grasses whose roots mat to form a close turf. Further progress may be halted by grazing pressure; if not, coarser grasses enter and give shelter to seedling shrubs and trees.

As mentioned briefly in Chapter 7, many plants are useful indicators of lime status. This is not, as is sometimes supposed, due to an absolute inability to grow in certain conditions, but because presence or absence of lime affects the powers of competition with other species. Thus, if competition is eliminated, most **calcicoles** (lime favouring) can be grown on lime deficient soils, and most **calcifuges** will grow on chalk soils. Many calcicolous plants have a much less restricted distribution on the continent. In Britain, at the extremes of their range, presence of chalk affects their powers of competition more than elsewhere.

Most agricultural grassland is 'neutral' (pH 6.0–6.5), based on clays, alluvia, and loams and supporting a mesophytic rather than xerophytic vegetation. Though the variety of species is less than on chalk, there is much to see and investigate. This is equally true of lawns and playing fields. Here light loving grasses are favoured and have much shorter and firmer blades than in the shade; many lawn weeds, like the dandelion, are affected in much the same way. The important factors of cutting, rolling, forking, manuring, herbicide or insecticide spraying, watering and even shading can be investigated on experimental patches; a control plot is of course essential. These procedures not only affect the species list, but they may encourage or inhibit particular growth habits such as side shoot development (tillering) in grasses. Many weeds are deep-rooted rosette plants which do not compete underground with more shallowly rooted plants. Other underground differences may have a bearing on the success of weeds; thus dandelion and dock with their large storage roots, and bind weed and couch grass with extensive rhizome systems are better able to survive ploughing or recover from weed spraying. Above ground, grasses and plants with dissected leaves are less affected by spraying than broad leaved plants. Many weeds (bracken, ragwort, buttercup) are distasteful and even harmful to animals so that selective grazing removes competition and the weeds are encouraged; this is particularly evident where grazing pressure is heavy as in rabbit warrens. Selective grazing may affect the flora in more subtle ways; without being actually distasteful, some plants may be less palatable than others, and the effect of this can only be discovered by careful investigation. Growth of nettles is associated with soils rich in both phosphate and nitrogen; these are found in sheltered areas where cattle and sheep congregate, or in the recognised latrine areas associated with rabbit burrows and badger setts. Other plants are affected by shelter, trampling and manuring, and it is easy to see how all three factors may occur together.

A less obvious example of shelter is provided by the ridges of ridge-and-furrow drainage systems, and the differences in water status of ridge and furrow are even more important. Thus plants favouring

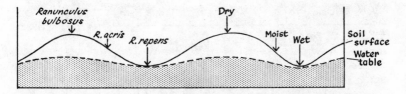

**Fig. 9.3**  Zonation of buttercups (*Ranunculus* sp.) on a ridge and furrow profile
(Courtesy of D. M. Barling)

drier conditions are found on the ridge and moisture loving species in the furrow – there may be a zonation of three buttercup species in relation to soil moisture (Fig. 9.3), and the field may show a banding of yellow from May to June because the species flower at different times.

Low-lying ground liable to flooding tends to support a number of characteristic moisture-preferring species, and forms **water-meadows.** Where the ground is perpetually wet the meadow grades into **marsh** with true hydrophytes (see Chapter 11).

The success of meadow plants and their powers of invasion can be measured by comparing the flora of permanent pasture with that of seeded pastures of different ages. Interesting comparisons can also be made between rough and cultivated grasslands; the rough, with taller coarser plants such as tor-grass, *Brachypodium pinnatum*, may show distinct stratification, whereas the closely grazed pasture will show less, yet may contain more species. The two may readily be compared on a golf course where the 'rough' meets the fairway, or at the edge of a playing-field. Plants such as dandelion, cat's ear (*Hypochaeris*), and plantain are favoured by mowing or grazing and are much less competitive if the grass is permitted to grow. Where myxomatosis in rabbits led to reduced grazing, the vegetation (unless checked by increasing the stock on the land) showed an increased trend towards the climatic climax. The early stages of such a change can be seen where land is acquired for building development and then neglected for a considerable time.

Tall herbs produce milder conditions favouring plants and animals not found in short turf (Fig. 9.4). In rough grassland it is usually possible to see small tunnels or runways made by the short-tailed vole *Microtus*, and less frequently by the bank vole *Clethrionomys*. Such runways often persist long after the individual animal has died, so it is necessary to look for more definite signs of life such as the bright green droppings, or heaps of cut grass stems. The vole has many enemies, particularly amongst the predatory birds – kestrels and buzzards by day, short-eared and barn owls by night. With the

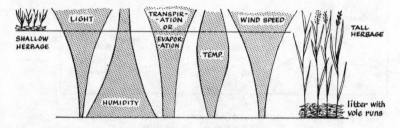

**Fig. 9.4**  Relationship between herbage height and habitat factors

decline of the rabbit, following myxomatosis, voles came to form the largest individual item in the diet of foxes (Fig. 9.5).

Stoats and weasels hunt in rough grass and hedgerows for small mammals including mice and rats which are more typical of woodland. Insectivores may be quite abundant in grassland. The mole lives in underground tunnels in permanent pasture and woodland, feeding on earthworms and insects and rarely showing itself. The hedgehog, as its name implies, requires some cover but the shrew is found in any low thick grass feeding mainly on earthworms, spiders and woodlice. Shrew remains are often found in owl pellets but the domestic cat, which also kills shrews, appears to find them distasteful and does not eat them.

The relationship between vegetation, small mammals, and inverte-brates is obviously very complex and needs following over long periods. In the vegetation alone, total frequency estimates for the main species may show that the pattern of dominance changes from one year to the next; this will be reflected by the changing abundance of associated animals, which however is more difficult to estimate. For the larger arthropods (grasshoppers, beetles, spiders) one-foot quadrat frames scattered in large numbers should be left for an hour or two before returning carefully, avoiding disturbance by shadow or ground vibration, to count the individuals in each square. Soil samples need to be treated in a variety of ways – Tullgren and Baermann funnels and flotation will each yield different smaller species, whilst breaking up by hand is sufficient to find the large insects.

The importance of the relationship between different species of plants, and between plants and insects, is well illustrated for ragwort, *Senecio jacobaea*, in a diagram due to Harper (Fig. 9.6).

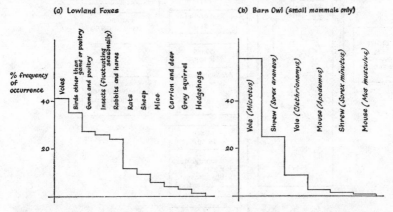

**Fig. 9.5** The importance of voles in the diet of predators since myxomatosis. [Data from (a) Lever, R. J. A. W. (1959), 'Diet of the Fox since Myxomatosis', *J. Anim. Ecol.*, **28**, p. 370, Table 4; (b) Gore, R. (1960), 'Prey of the Barn Owl *Tyto alba*', *Cheltenham G. S. Bio. Rep.*, No. 4.]

Ragwort competes with grass and other herbs for its basic needs of light, water, and soil nutrients. If it is removed, the equilibrium will alter – probably in favour of grasses, but possibly in favour of some other weed (ragwort draws water from lower in the soil than many grasses so it may well be replaced by a weed with similar root pattern). Apart from the direct effect on the vegetation, an enormous number of animals and parasitic plants depend on ragwort and the effect on them may lead to other changes in the community. Such patterns should be worked out for other species of plants, bearing in mind the fact that relationships may change seasonally with the stages of development. Thus seeds and fruits are food for animals which may show no interest in other stages. Early seedlings may be quite different from later stages, as in the case of gorse where the soft trifoliate leaves are quite unlike the hard woody spines that represent the later leaves.

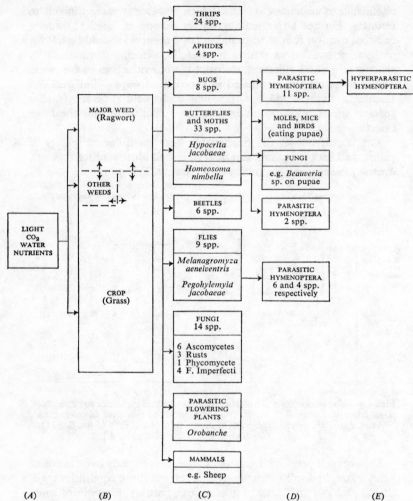

(A)              (B)                  (C)                    (D)              (E)

The columns denote: (A) plant growth requirements; (B) weed-crop relation-ships; (C) organisms attacking ragwort; (D) parasites and predators; and (E) their hyperparasites. The numbers of species are those so far recorded in the U.K., but the list is far from complete, particularly for parasites and hyper-parasites. The arrows in (B) indicate the changes in composition of a weedy crop (ragwort-infested pasture, in this instance) which may follow changes induced in the food chain: expansion or contraction of the ragwort population is relative to that of the crop or other weeds. Herbicidal treatment of ragwort may be followed by re-invasion of other weeds or expansion of the crop, or both, and all the organisms in the food chain are affected

**Fig. 9.6** Diagrammatic food chain of a weed exemplified by ragwort, *Senecio jacobæa*. [From Harper (1957), Ecological Aspects of Weed Control, *Outlook on Agriculture*, **I**, No. 5.]

Do any animals feed on this seedling stage? Only careful observation and recording will provide the answer.

Just as each plant provides a microhabitat, so the droppings of farm and other animals form an interesting basis for study. Cattle dung provides the basis of a microseral succession lasting only as long as it takes the droppings to dry. Obligate breeders on dung are followed by dung beetles and other coprophagous (dung feeding) forms with all their associated predators and parasites. The droppings of wild animals, if analysed (see Chapter 15), can provide useful information about their diet.

The problems to be studied vary with the nature of the grassland. Acid grasslands are found on sandstones and shales where lime is absent and leaching is considerable. The resulting absence of bacterial decay may lead to the formation of a humus which returns little nutrient to the soil. Much of the semi-natural acid grassland is dominated by fescues (*Festuca* sp.) and bents (*Agrostis* sp.). After burning, however, the pattern may be modified so that wavy hair grass (*Deschampsia flexuosa*) or creeping soft grass (*Holcus mollis*) becomes dominant. Opportunities for studying succession following burning should not be neglected. Bracken or gorse scrub are common invaders of hill grassland and show considerable powers of recovery after burning; this is particularly true of bracken which has extensive underground rhizomes. In damper conditions where acid peat accumulates, mat grass (*Nardus stricta*) communities are common, and where the soil is perpetually moist, purple moor grass (*Molinia caerulea*) may dominate. As the name suggests, the latter grass is commonly associated with moorland which may in fact develop from acid grassland; heaths and moorland are considered in the next chapter.

# 10 Heath, Moorland and Bog

To most people, 'heath and moorland' conjures up an image of wide undulating country exposed to sun, rain and wind, with a vegetation dominated by the low shrubs of heather (*Erica*), ling (*Calluna*) and possibly gorse (*Ulex*). Whether it is dry or wet underfoot, the general picture is much the same. All three of these plants show marked adaptations to prevent excessive water loss from the leaves, and this underlines a major characteristic of these habitats – that plants have difficulty in obtaining sufficient water for their needs. The reasons for water shortage differ between heath and moorland. Exposure to sun and wind coupled with a porous freely drained sand or gravel soil results in a genuine water lack in heathland. In places, however, leaching may produce a **hardpan** which restricts drainage and results in **wet heath**. On moorlands, the water shortage is physiological; the wet windswept soil is always cold, and this restricts water uptake.

There are no fixed rules for defining the term 'moorland', so that where one type of situation grades into another, the terms **moorland** and **wet-heath** may be interchanged.

An example of such a transition from dry to wet conditions is shown in Fig. 10.1. Here, leaching of the Devonian sandstone has resulted in a typical podsol with a shallow peaty topsoil over a hardpan. On the drier high ground, the vegetation is dominated by ling or bracken (*Pteridium*) together with bell-heather and gorse. The older ling has died back leaving spaces soon occupied by bell-heather, mosses and lichens. Water is retained on the shallow lower slope by the bog moss *Sphagnum* which forms a layer of waterlogged peat. Near the edge of this, ling and bell-heather are replaced by cross-leaved heath (*Erica tetralix*). Growing in the peat and around the pool are various kinds of rush (*Juncus* sp.), bog asphodel (*Narthecium ossifragum*), sundew (*Drosera*), and the two sedges 'deer-grass' (*Trichophorum caespitosum*) and 'cotton-grass' (*Eriophorum angustifolium*). In other parts of the same moor where *Sphagnum* has accumulated in damp valleys to form slightly raised blanket-bog, the black bog rush (*Schoenus nigricans*) and butterwort (*Pinguicula*) are found.

In moorland with higher rainfall, bilberry (*Vaccinium*) often grows amongst the ling, sometimes becoming dominant when given the chance. In deep waterlogged peat in very exposed conditions (as are found on high ground in the Peak District and parts of Scotland), cotton-grass or deer-grass may become dominant.

Where moor or heathland is pastured and trampled, the grasses increase to give a predominantly grassland vegetation. The principal grasses are the bents (*Agrostis tenuis, A. stolonifera, A. canina* generally; *A. setacea* in the south west), wavy hair grass (*Deschampsia flexuosa*) and sheep's fescue (*Festuca ovina*) on the dry heaths, purple moor grass (*Molinia caerulea*) on wet moorlands and mat-grass (*Nardus*

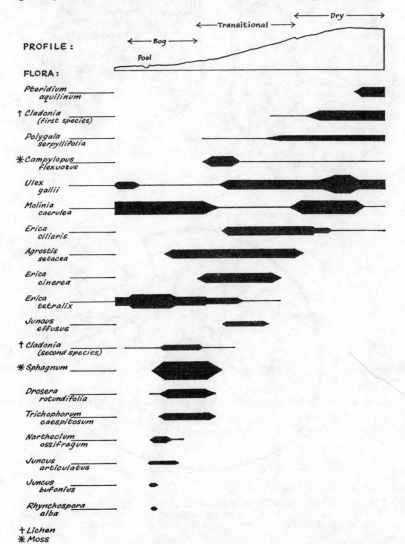

**Fig. 10.1** Profile transect, Woodbury Common, Devon. Thickness of line indicates abundance

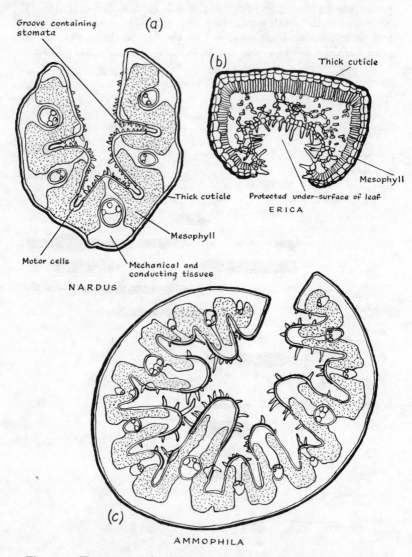

**Fig. 10.2** Transverse sections of xeromorphic leaves. (a) Mat grass (*Nardus*). (b) Bell heather (*Erica*). (c) Marram grass (*Ammophila*—distribution of tissues as for Nardus). In each case stomata are restricted to protected surfaces

*stricta*) on the better drained areas of both heath and moor. *Nardus* forms tussocks which accumulate humus and raise the soil level.

Disturbed areas, burned patches and bogs all provide opportunities for investigation by means of transects aligned to cut across the different phases of vegetation.

Many heath and moorland plants show adaptations which meet the difficulties resulting from water and nutrient shortages; those which assist survival in water shortage are said to be **xeromorphic**. They include:

(1) THICK WAXY CUTICLE. This limits water loss to transpiration through the stomata. A shiny surface of this kind also reduces light and heat absorption and therefore lessens the need for the cooling effect of transpiration. In many species, the cuticle thickness is greater in plants grown in the open than where there is some shade.

(2) PROTECTED STOMATA. There are various ways in which air movement near stomata is restricted. Stomata may be enclosed by leaf rolling or folding, as seen in many grasses (*Deschampsia, Festuca, Nardus*; also marram grass, *Ammophila* – see also Chapter 13), or protected by hairs, or sunken in pits or grooves (*Calluna* and *Erica*) (Fig. 10.2).

This protection lengthens the diffusion path from the saturated air near the leaf mesophyll to air at atmospheric humidity, and so reduces the rate of water loss.

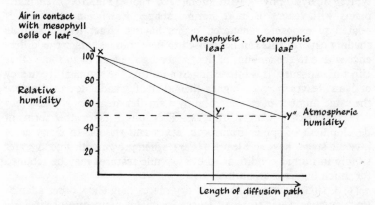

**Fig. 10.3** Relationship between diffusion path and diffusion rate. XY′ and XY″ represent diffusion gradients and hence diffusion rates (rate is faster for steeper gradient)

(3) REDUCED LEAF SURFACE. The ratio of leaf surface to volume is reduced by the leaf becoming thickened (*Calluna, Erica*) or even circular in cross section (Fig. 10.4). In some cases leaves may be reduced to scales (broom, *Sarothamnus*) or spines (gorse), so that

other structures like petiole or stem, with fewer stomata and more woody tissue, become the main photosynthetic organs. The various species of *Juncus* are particularly interesting examples, because they combine the xeromorphic photosynthetic stem with aerenchyma tissue

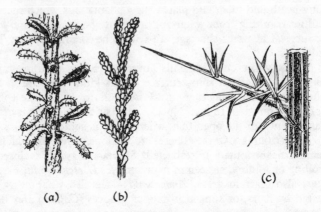

(a)          (b)                                    (c)

**Fig. 10.4** Reduction of leaf surface in (*a*) bell heather (*Erica*); (*b*) ling (*Calluna*); (*c*) gorse (*Ulex*). In gorse leaves and stems are modified as spines; both are photosynthetic

typical of hydrophytes with submerged roots (Fig. 10.5). In many plants with leaves of more normal shape (*Vaccinium*), leaf size is related to exposure, sheltered plants having larger leaves. Simple shading experiments can be devised to determine whether this difference is due to the amount of light reaching the leaf, or to some other effect of exposure. It is interesting to note that the stomatal frequency of 'sun' leaves may be higher than that of 'shade' leaves, because the same number of stomata occupy a smaller area.

(4) INCREASED WOODY TISSUES. This often takes the form of development of spines from both leaves and shoots. In many cases juvenile stages have soft leaves (*Ulex*); when grown with a good water supply in humid conditions, these juvenile features may be retained for much longer than normal.

(5) OTHER METHODS. Plants in dry places may show other adaptations such as water storage tissues. Succulent plants are particularly typical of 'deserts' (e.g. stonecrops growing on nearly bare rock), and where the salt content of soil water impedes water uptake as in salt marsh or on sea cliffs. Some plants are drought avoiders – either annual plants surviving difficult periods as seed, or perennial plants which die back seasonally. Many moorland plants (*Pteridium*, *Molinia*, *Trichophorum*) die back in this way.

Not all moorland plants show obvious xeromorphic features and their ability to survive is worth investigating. In some cases (milkwort, *Polygala*) the shelter given by larger plants is obviously important; in others (the mosses *Hypnum* and *Polytrichum*, the lichen *Cladonia*) ability to recover from almost complete drying is important. The particularly extensive root system of purple moor grass (*Molinia*) maintains

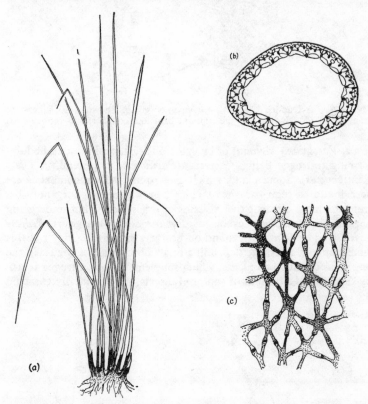

**Fig. 10.5** The common rush (*Juncus effusus*). (*a*) Whole plant with photosynthetic stem, and no leaves. (*b*) T.S. of stem showing thick cuticle and abundant mechanical tissues. (*c*) Aerenchyma cells from centre of stem; the air spaces extend into the roots

a good rate of water uptake and eliminates the need for special transpiration checks.

On moor and heath little true humus is formed because bacteria are inhibited by either the dryness of heathland or the cold waterlogged conditions of moors, and this results in a shortage of nutrients, particularly nitrates. This tends to limit the flora to plants whose requirements are small (mosses, lichens), and those which can obtain nutrients through mycorrhizal associations with soil fungi (*Calluna*,

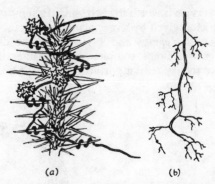

**Fig. 10.6** (a) Dodder (*Cuscuta epithymum*) on gorse. The stem is deliberately exaggerated. (b) Nodules on the roots of birdsfoot trefoil (*Lotus corniculatus*)

*Erica, Vaccinium, Molinia*) or by symbiotic relationships with nodule-forming nitrogen-fixing bacteria (*Papilionaceae* generally; *Ulex, Sarothamnus*). Some plants which grow quite well on good soils are here found as semi-parasites taking their water and inorganic salts from the more efficient root systems of other plants, but undergoing normal photosynthesis (eyebright, *Euphrasia*, and lousewort, *Pedicularis*). Dodder (*Cuscuta*), found on heathers and gorse (Fig. 10.6), is an example of a parasite depending on its host for all its needs. In the bog areas carnivorous plants, which supplement their nitrogen intake by digesting the protein of captured insects, may be locally common

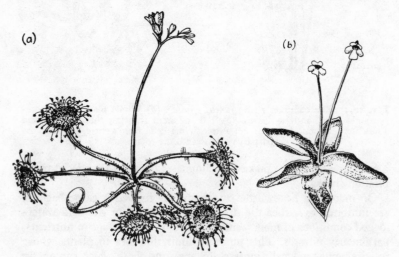

**Fig. 10.7** Carnivorous plants. (a) Sundew (*Drosera rotundifolia*). (b) Butterwort (*Pinguicula vulgaris*)

(sundew, *Drosera*; butterwort, *Pinguicula*; and, occasionally in bog pools, bladderwort, *Utricularia*).

In the exposed, acid, and often waterlogged conditions, the soil and ground fauna is more sparse than on more temperate grasslands, and many of the mammals and birds have only ventured there from the shelter of woodland or hedges to hunt food. However, heather and bracken provide shelter for various animals and plants. Tree seedlings may enter, so that pine or birch may supersede the heather if not checked by grazing (sheep, rabbits) or by regular burning. Moors are often preserved by controlled burning to encourage the nesting of game birds. The animal and plant successions following this treatment are well worth investigating.

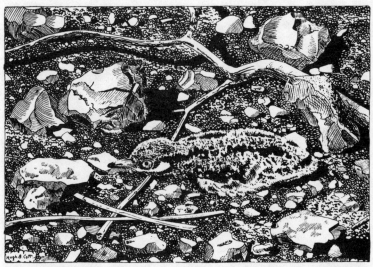

**Fig. 10.8**  Squatting attitude and disruptive markings conceal the outlines of a young Stone-Curlew; its main body colour matches the background well. (From H. B. Cott, *Adaptive Coloration in Animals*, Methuen)

In general, except where there is shelter, only the stronger flying insects such as bees and wasps are common. However, in shelter many ground living insects may be found, together with beetles, spiders, lizards, snakes, and a variety of birds which prey on them all. Many escape notice because of **cryptic colouring** (Fig. 10.8) which merges with the background, or **disruptive colouring** which distracts the eye and hinders recognition of familiar outlines. In birds, this camouflage usually extends to the eggs as well. Predatory birds (hawks, owls) and mammals (stoats, weasels) are found wherever there are seed and fruit eating birds and small mammals. In some areas (e.g. Dartmoor) minks which have escaped from fur farms have established themselves successfully near moorland streams. Little is known of their present way of life.

The peaty waters of moorland pools and streams support a very limited flora and fauna. Suggestions for the study of freshwater life are given in the next chapter.

# 11 Freshwater

Freshwater habitats are convenient and rewarding to study because it is usually easy to see how environmental factors influence the composition of the community and the structure of individuals. In addition, freshwater species are fewer and more readily identified than land organisms and most can be kept alive in the laboratory for further observation.

## The study of different habitats

The best introduction to freshwater ecology is to collect and identify plants and animals from a few distinct habitats and so acquire a grounding in the art of recognizing species and their adaptations to particular conditions.

For convenience, freshwater may be divided into still and flowing water habitats. As a first step in comparing these, a pond could be compared with a stream by collecting animals and plants and noting which species occur in both pond and stream, and which are limited to one only. Later, different habitats within the one pond or stream can be compared.

Sometimes it is clear how structure or habits limit distribution; thus duckweed and pond skaters are obviously not suited to life in a swift stream because they would be swept away. In other species it may be more difficult to explain distribution, so it is always important to record habitat factors and look for signs of competition and other relationships between species which may help to determine the character of the community.

### *Ponds and lakes*

There is no fundamental difference between a pond and a lake except size; disused canals and ditches may also provide the same conditions.

In still waters many important factors vary with depth (Fig. 11.1). Temperature, light intensity, oxygen tension and pH tend to have higher values near the surface. Carbon dioxide tension, quantities of inorganic substances and organic debris all tend to increase towards the bottom. These graded differences may help to explain why some animals are most abundant at particular depths. Bottom habitats tend also to be related to depth, and form a graded series from

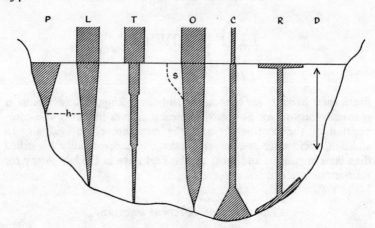

**Fig. 11.1** Relationship between habitat factors and depth in still waters. P=plants, L=light, T=temperature, O=oxygen, C=$CO_2$, R=organic débris, D=distance, h=limit for photosynthesis, S=supersaturation with oxygen in daytime owing to photosynthesis

shallow to deep water. This pattern is usually modified by other factors such as the type of shore material, exposure to wave action or shade from overhanging trees, so that the habitats form a **mosaic** (Fig. 11.2).

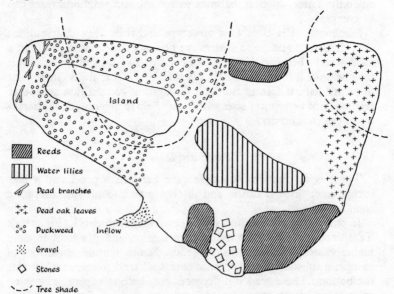

**Fig. 11.2** The mosaic of habitats in the upper three inches of a large pond. Wind and wave action may alter the position of the duckweed

Green plants are limited by the depth to which useful light penetrates so that rooted plants are confined to water a metre or so deep – usually near the shore. Some species can survive with less light than others and so may live in deeper water. The nature of the bottom is also important in determining which plants can become established; factors involved include texture (stones, sand or silt), the quantity of organic matter and the rate new material is deposited by inflowing streams or from organisms dying in the lake itself.

Lakes and ponds are excellent places to study plant succession. Submerged colonisers may include various species of pondweed *Potamogeton*, water milfoil *Myriophyllum*, Canadian pondweed *Elodea* and the stoneworts *Chara* and *Nitella*. The remains of these plants tend to stabilize the bottom by binding together loose materials; they also add to its depth and organic content. As the nature and depth of the bed changes, new species can enter and add more organic matter. In this way a definite succession of plants progressively changing the habitat commences with submerged forms, and gives way first to plants with floating leaves (e.g. *Potamogeton natans*, waterlily *Nymphaea*) and then to types with emergent aerial leaves or stems like reeds, sedges and horsetails (Fig. 11.3).

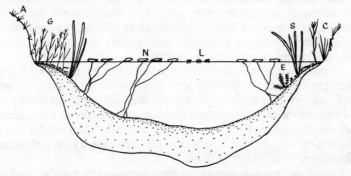

**Fig. 11.3** Plant zonation in a disused canal. On the bank is *Agrostis* (A); the reed swamp contains *Carex* (C), *Glyceria* (G) and *Sparganium* (S); *Nymphaea* (N) and *Lemna* (L) float on the open water; *Elodea* (E) is submerged

The succession continues on to land; in acid regions peat moss *Sphagnum* accumulates to form bog, whilst where it is less acid various rushes, grasses and sedges build up the reed swamp to form fen. Willow and alder may then enter the succession and eventually give rise to fen woodland or carr (Fig. 11.4). It is rare, however, to find the complete sere. An outline map of the vegetation and one or more profile transects are suitable for recording the succession. Since the particular species involved may vary, it is interesting to compare the successions in different water habitats. The vegetation seldom conforms exactly to the above pattern because zonation is often disturbed by shading, streams, and biotic interference such as cutting or grazing.

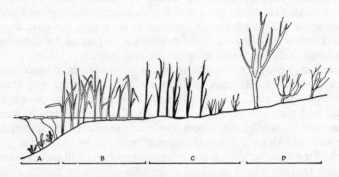

**Fig. 11.4** Diagram of a hydrosere. (*a*) Submerged and floating-leaved communities: *Potamogeton, Elodea, Nymphaea*. (*b*) Reed swamp: *Scirpus*. (*c*) Fen: *Phragmites, Carex, Filipendula*. (*d*) Carr: *Alnus, Salix*. (Transect taken at Esthwaite Water.)

Free-floating plants like the fern *Azolla* and duckweed *Lemna* are found mainly in sheltered bays and ponds where there is little wave action. An interesting study could be made relating the distribution of different species to influences such as lake size, shelter, animal life, and the amount of organic remains.

In open water it is worthwhile to take regular samples of plant plankton throughout the year and note the changes in species and their relative abundance. Desmids are generally common during the cold months whilst diatoms and filamentous types abound in summer. All are affected by light, temperature, and pH, but a sudden fall in population may be due to depleted reserves of inorganic substances like silicon or to attack by fungi.

Freshwater animals may be grouped according to whether they are found at the surface, in mid-water, burrowing in the substratum, or on plants, stones and dead leaves.

Animals found at the surface are supported there by surface tension. Those living on top of the film, like the pondskater *Gerris* and the water measurer *Hydrometra*, are usually carnivorous and often eat terrestrial insects that have fallen into the water. The underside of the film is sometimes temporarily inhabited by planarians, snails like *Limnaea pereger* and *Planorbis*, mosquito larvae, and other organisms which normally live elsewhere.

Animals found in mid-water include weakly swimming plankton, strong swimmers like sticklebacks, and creatures like water boatmen and beetles, which pass through on the way to breathe at the surface.

Water fleas (cladocerans) and copepods are common in the plankton, and both groups feed by filtering small organic particles and algae from the water, though some are actively predaceous. Smaller plankton includes the rotifers and many kinds of protozoans, particularly ciliates. Some species in the plankton are most abundant over deep water, some near rooted plants, and others close to large amounts of organic debris. This uneven distribution and the seasonal changes in species and abundance can be studied by careful sampling.

The lake bottom is a patchwork of microhabitats such as dead leaves, stones and rooted plants. Each has a slightly different association of species although in ponds some animals like *Limnaea pereger* and *Asellus* are found almost everywhere. Species also vary with water depth, and between the zones of plant succession around the margin of the lake. Is animal succession a direct consequence of plant succession? It should be possible to find out.

The type of shore is also important. Thus a sheltered shore with a bottom abounding in organic matter and rooted plants has a fauna which is very different from that of an exposed rocky shore. Sedges and reeds reduce water movement to a minimum, so that an overgrown bay of a large lake may resemble a pond in this respect. On a rocky shore wave action causes turbulence and many animals characteristic of streams, like the river limpet *Ancylastrum*, may occur. If you have access to a large lake, you may learn much by collecting and listing species from different shores and comparing their abundance. Where a stream enters, the changes are striking.

### Streams and rivers

The most important factor in streams and rivers is the current by which silt, pebbles, and even large rocks may be carried downstream. As the current slows, first stones and then gravel and progressively smaller particles are deposited. Because water speed varies in different parts of the stream and is slower near the edges, flowing waters, like lakes, are also broken up into a mosaic of habitats – in this case based on fairly uniform zones of stones, gravel, sand and mud. Any study of a section of stream should include an account of this habitat mosaic.

Water movement also influences whether organisms are found at the surface, in mid-water, or on the bottom. Free floating organisms cannot maintain their positions in the main stream so that the only plankton is that carried out of lakes in the current. Surface dwellers like the water cricket *Velia* are found only in still pools and backwaters.

The river bed is the most heavily populated because organisms can shelter or attach themselves firmly so that they are not swept away. The species vary according to the current and the bed materials. Nymphs of one stonefly, *Dinocras cephalotes*, are common in rivers with firm beds of angular stones, but where the rocks are rounded and less stable another stonefly, *Perla bipunctata*, is more characteristic. Mosses and the water crowfoot *Ranunculus fluitans* are common in fast flowing streams whilst the pondweeds *Potamogeton* and *Elodea* characterise still waters. The scouring effect of periodic flooding may keep some areas quite bare of plant life.

A stream should be studied both by quadrats and profile transects made along and across its course to show the nature of the bed and the distribution of plants and animals (Fig. 11.5). Many creatures hide

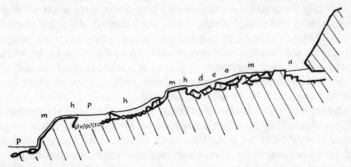

**Fig. 11.5**  Profile of stream emerging from a limestone fissure. Mosses (*m*) are abundant on the waterfalls and near the spring. The flatworm *Planaria* (=*Crenobia*) *alpina* (*a*) is associated with the spring. The net-spinning caddis *Hydropsyche* (*h*) occurs in swift water whereas *Plectrocnemia* (*p*) spins in the pools. The mayfly and stonefly larvae *Ecdyonurus*, *Baetis* (*e*) and *Dinocras* (*d*) occur in moderate currents where there are loose rocks. (From a survey at Fairy Hole Cave, Co. Durham.)

under stones and may be caught by a net placed downstream. When the stone is lifted, the animals are swept into the net. Because animals tend to be swept away like this, collection of fauna from a quadrat frame placed on the stream bed should begin at the downstream border (Fig. 3.10). Algae may be collected by scraping samples off rocks or by hanging glass slides in the stream. They settle and grow on the glass.

Temperature, light intensity and organic litter must also be taken into account when comparing collections from different stations along a water course. Thus, if the stream passes through a wood, light intensity is reduced and dead leaves accumulate; this hinders plant growth but encourages scavengers like the water louse *Asellus*.

Collections made in rivers flowing from contrasting geological areas should also be compared; some organisms like crustaceans and molluscs tend to be more plentiful in hard waters.

## Pollution

Many unwanted materials are dumped in inland waters. These range from old bicycles and tin cans in ponds to sewage, poisons, detergents, and hot water discharged into rivers. Most rivers are used to remove industrial wastes and sewage – processes which both reduce their aesthetic and recreational value, and limit their usefulness as sources of domestic water supply. Although natural ecosystems can degrade and eliminate small amounts of some pollutants, discharged wastes often exceed the natural limit and include non-degradable materials. In addition to the toxic effects of pesticides or heavy metal ions, effluents and rubbish modify habitat factors such as pH, oxygen content, hardness of the water, and the substratum. Each kind of pollution affects the freshwater community in a different way; some species are more susceptible to poisonous outflows from chemical works and zinc mines, whilst others succumb easily to inert waterborne particles like china clay or coal dust which may interfere with breathing or feeding mechanisms.

Pollution by sewage has marked effects on fish populations, and untreated sewage may contaminate the water with pathogenic microorganisms. Quite small quantities of sewage may change the proportions of different organisms in a stream – possibly by stimulating algal growth. The effects become striking when the water is deoxygenated by excessive decaying material; bacteria which decompose organic material make heavy demands on the available oxygen. When oxygen demand exceeds supply, anaerobic micro-organisms multiply in their turn. White and brown masses of 'sewage fungus' (an association of fungi, bacteria, and sessile ciliates) coat the river bed. Algae and other green plants are adversely affected by the heavy silting, reduced light intensity, and the lack of oxygen. Animals are most affected by oxygen lack, but silting and the disappearance of plants are also important. Organic pollution is not always man-made – deoxygenation and sewage fungus often result from natural accumulation of dead leaves or cattle dung.

Downstream from a heavily polluted region, the oxygen level gradually increases (Fig. 11.6). Typical first colonizers are *Asellus*, chironomid larvae, and worms like *Tubifex*. As the stream recovers, more species are found and a sample collected at any one point will fairly reliably indicate the degree to which the river is still polluted. Such collections are used to make rapid assessments of river quality because they reflect conditions over a period of days or weeks. In contrast, single samples taken for chemical analysis indicate only the immediate conditions and may therefore fail to show up periodic outbursts of more serious pollution.

The breakdown of sewage releases a rich supply of mineral salts so that as conditions improve downstream a profusion of algae appears, particularly blanket weed, *Cladophora*, which forms thick floating mats with tangled bright green filaments often several metres long. A similar effect sometimes results from excess nitrates, whilst phosphates from detergents may also cause excessive algal growth. In lakes, this extra growth may so increase the decaying material at the bottom that oxygen depletion of the lower layers causes impoverishment of the aquatic community. Lakes which do not accumulate enough organic debris to cause oxygen depletion are said to be *oligotrophic*. Naturally fertile lakes where death of abundant phytoplankton eventually produces oxygen lack at

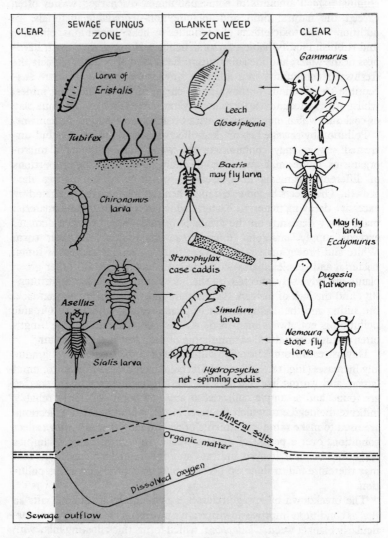

**Fig. 11.6** Diagram illustrating the ecological recovery downstream from a source of organic pollution. Invertebrates characteristic of each zone are illustrated, though some overlap may occur as indicated by the arrows. As recovery proceeds the diversity of species increases and this diversity may be used as a biological index of water quality

lower levels are described as *eutrophic*. The changes brought about by human enrichment or *eutrophication* of lake waters with nutrient materials must therefore depend on the original status of the lake.

Thermal pollution results from river water being used for industrial cooling purposes – mainly by power stations. Temperature may be raised by as much as 10–15°C and remain warmer than normal for some miles downstream. The raised temperature accelerates organic decay and increases biochemical oxygen demand (B.O.D.). At the same time, warmer water can carry less oxygen so that the more active animals are soon affected and possibly eliminated altogether. The development of eggs and larvae may be blocked or accelerated, and hatching and metamorphosis may be prevented or occur at the wrong time of year.

The effects of pollution may be studied by:

1. Sampling at stations upstream, at the outflow, and downstream to find which species are present and in what numbers.
2. Recording changes in the community as they occur, where it is suspected that new pollution is occurring.
3. Studying recolonization where pollution has been stopped.

Wherever possible, records of community changes should be accompanied by chemical and B.O.D. determinations and by discreet enquiries at the polluting source (sewage works, factory or farm) so that the exact nature and concentration of pollutants may be established.

It is always useful to have some knowledge of the processes producing effluent. In a sewage works, much of the degradable matter is removed, first in settling tanks, and then by trickling the remaining liquid over a filter bed of gravel and stones – which provide a large well aerated surface where decay bacteria break down organic material. These bacteria are kept in check by an animal community which thus prevents the filter bed becoming choked. Pollution occurs when a town population outgrows the capacity of its sewage works so that treatment is incomplete. However, even where treatment is complete, the production of nutrient salts in the process often gives rise to eutrophication.

### Food relationships

A convenient introduction to the feeding relationships of water organisms is to work out a few simple food chains. Because their inhabitants are easy to keep alive in the laboratory, ponds and lakes are the best places for such studies.

A typical food chain leads from microscopic algae, through planktonic crustacea and sticklebacks to water birds. Dead organisms and pieces of food wasted during feeding form the diet of bacteria and scavengers. Many kinds of parasite also occur. The pyramid of numbers (Fig. 2.1, p. 8) can be well illustrated by quantitative studies in ponds. When several food chains are known it will be seen that they are interlinked and part of the food web of the community can be worked out. When doing this it must be remembered that many freshwater organisms are abundant for a few months only so that some links in the chain may disappear and new ones come into being. Thus frog tadpoles are a food source for predators only in spring and early summer so the food web must alter with the seasons.

Some ponds support more life than others. As all animals depend ultimately on plants, a useful way of comparing the **productivities** of different ponds is to make quantitative estimates of plankton using a microscope and standard samples of pond water (e.g. in a haemocytometer cell used for counting blood corpuscles). The two samples may first need to be concentrated in a centrifuge. If similar volumes are centrifuged at the same time, the numbers recorded for particular plankton species in the concentrates will be related to actual abundance in exactly the same way.

Most food chains in lakes start with plant plankton, but in ponds and streams organic matter such as dead leaves from land plants may be equally important.

## The study of adaptation

There are many ways in which living things have become adapted to life in freshwater habitats. Where quite different species meet the same problems, they adapt to them in ways that are often strikingly similar. In the following pages a few examples of adaptation are quoted as a guide only; the best way to learn about adaptation is by direct observation.

### Maintaining ecological position

Both structure and behaviour help to keep the animal within the habitat to which it is adapted. Sometimes they keep the animal in the same area, but if the habitat changes the animals may have to move to remain in the same effective environment. Putting it in another way, they have to move *to maintain their ecological position*. If the animal itself undergoes changes, it may have to move to a quite different kind of habitat as when dragonfly nymphs change into winged adults.

Stream animals show striking adaptations enabling them to keep their position in a water current (Fig. 11.7). Some attach themselves to stationary objects by means of adhesive structures like the suckers of a leech, the foot of a water snail, or the under surface of a planarian. Other animals like mayfly and stonefly nymphs cling on with hooked claws. The freshwater sponges and caddis larvae which stick their stony cases to rocks make more permanent attachments.

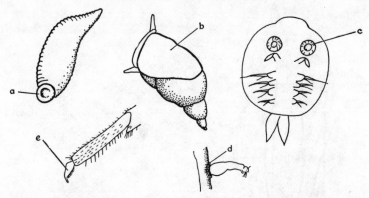

**Fig. 11.7** Devices for attachment: (*a*) sucker of leech (*Glossosiphonia*); (*b*) foot of snail (*Limnaea*); (*c*) suckers of fish louse (*Argulus*); (*d*) silk pad to which the *Simulium* larva hooks itself; (*e*) claw of mayfly nymph (*Baetis*)

Having a streamlined shape reduces the effect of the current (Fig. 11.8). Plants in swift streams are cushion shaped (mosses) or have strong flexible stems. In many aquatic plants the leaf shape varies according to water speed and whether leaves are submerged or not (Fig. 11.9). The flattened form of many stream animals enables them to avoid the current by hiding in crevices or under stones like *Gammarus*,

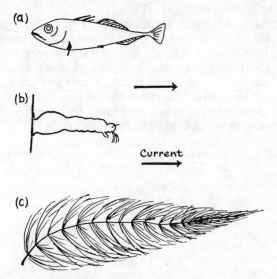

**Fig. 11.8** Streamlined shape of some aquatic organisms. The blunt end faces upstream: this is (*a*) the anterior end of the stickleback, but (*b*) the posterior end of the blackfly larva (*Simulium*). The flexibility of (*c*) *Myriophyllum* allows the current to form it into a streamlined shape

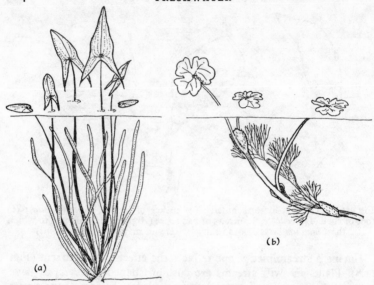

**Fig. 11.9** Leaf form and submergence. (*a*) Submerged, floating and aerial leaves of arrow-head (*Sagittaria*). (*b*) Aerial and submerged leaves of water crowfoot (*Ranunculus*)

or by clinging tightly in the slower moving water close to the stone surface like *Ecdyonurus* (Fig. 11.10).

Structural modifications are accompanied by adaptive behaviour like creeping into crevices in response to contact or by moving away from the light. Some planktonic animals have flotation devices like the spines on certain water fleas and the hydrostatic air sacs in the phantom larva *Chaoborus* (Fig. 11.11). These adaptations may be accompanied by a vertical migration – many rise closer to the surface at night. These movements may be studied by collecting samples at different times or by experiment in the laboratory.

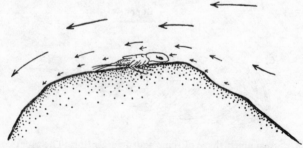

**Fig. 11.10** The flat form of the mayfly nymph (*Ecdyonurus*) keeps it in the low current 'boundary zone' near the rock surface. Length of arrow indicates current speed

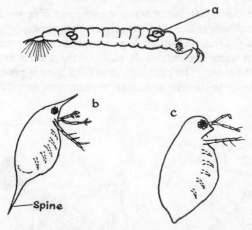

**Fig. 11.11**   Flotation devices: (*a*) air sacs of phantom larva (*Chaoborus*); (*b*) water fleas (*Daphnia*) with and (*c*) without long spines

## Feeding adaptations

Freshwater animals may have adaptations enabling them to feed as filterers, carnivores, scavengers or herbivores.

In still waters filter feeders must create their own current to bring in food. Suspensions of small particles like yeast cells or fibrin stained with congo red may be used to see how food is drawn in. In streams, food particles are brought to filter feeders by the current. Net-spinning caddis larvae build silk nets and explore them at intervals for trapped food, while *Simulium* catches small particles on large mouth brushes (Fig. 11.12). An improvised trough for running water can be a great help for observing stream dwellers.

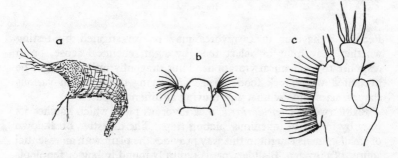

**Fig. 11.12**   Filtering devices: (*a*) net of net-spinning caddis; (*b*) mouth bristles of *Simulium*; (*c*) limb of water flea with food-filtering fringe

Carnivores have piercing or biting mouthparts. The water boatman *Notonecta* and beetles like *Dytiscus* actively seek their prey. Dragonfly nymphs and the water-scorpion *Nepa* usually lie in wait ready to grab their victims using the extensible mask of the former and raptorial forelegs of the latter. The phantom larva *Chaoborus* is peculiar among insects in using prehensile antennae to secure its prey (Fig. 11.13).

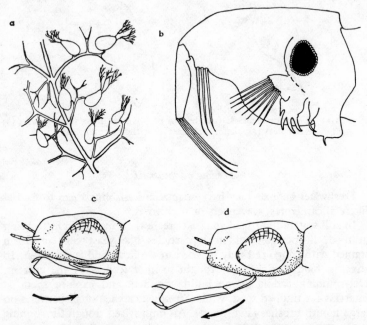

**Fig. 11.13**  Predatory devices: (*a*) *Utricularia*, bladderwort, part of an under-water leaf bearing animal-trapping bladders; (*b*) head of *Chaoborus* showing prehensile antennae; (*c*) resting and (*d*) capture positions of the 'mask' of a dragonfly

Feeding behaviour in carnivores may be investigated by testing whether they hunt or select food by sight, chemical sense, or in response to mechanical vibrations. Thus dragonfly nymphs might be presented with mock food objects at varying distances, in various shapes, sizes and colours, and moving at different speeds.

Bladderwort (*Utricularia*) is a carnivorous plant which catches its prey by means of ingenious suction traps. The digestive breakdown of small animals caught in this way provides the plant with an essential source of nitrogen. Bladderwort is typically found in bog or fen pools where organic remains do not decay quickly.

Little is known of the food preferences of scavengers like corixids or of herbivores such as water snails, so there are plenty of opportunities for making new discoveries. Do snails live on the tissues of rooted plants or merely browse on algae which are epiphytic on the plants ?

Food choice may change as the animal gets older; thus a newly hatched *Dytiscus* larva is not likely to attack the same size prey as it will when full grown. In cases like these, gut contents may provide information about food habits.

### Modifications aiding respiration

Submerged parts of plants cannot always obtain enough oxygen from their surroundings. In most water plants there is an extensive and characteristic tissue called *aerenchyma* which has large air spaces through which oxygen produced during photosynthesis can diffuse to all parts of root, stem and leaf. (Fig. 10.5c, p. 149).

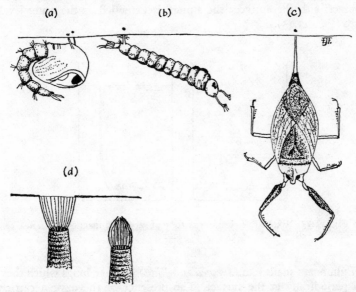

(a)        (b)        (c)

(d)

**Fig. 11.14** Adaptations to surface breathing and flotation. Respiratory siphons of (a) pupa and (b) larva of *Culex*. (c) *Nepa*, the water scorpion. (d) Hydrofuge hairs opening at the surface and protecting the spiracles when submerged, as a result of surface tension (*Stratiomys* larva—after Wigglesworth)

Some aquatic animals rely on dissolved oxygen for their needs whilst others obtain oxygen direct from the air. Turbulence from wave action or strong currents impedes renewal of air supplies at the surface so that air breathers are found mostly in slow-flowing and sheltered waters. Many are insects which may take down a film or bubble of air on their body surface, or merely renew the air in their tracheal systems. The spiracles guarding the openings to the tracheae are usually fringed with hydrofuge (non-wetting) hairs which prevent water flooding in. In *Nepa* the spiracles are also set at the end of a long respiratory siphon. Both siphon and hairs help to break through the surface film (Fig. 11.14). *Hydrophilus*, the silver water beetle, obtains air in an odd way by using a channel between its head and modified antennae.

Insects which take down air bubbles may carry them under the elytra (*Dytiscus*), on the hind end (*Gyrinus*), or over much of the body surface (*Hydrophilus* and *Notonecta*). An air bubble may act as a temporary gill because as oxygen is used up the store is partly replenished by diffusion from the surrounding water. This principle is also used in the underwater air stores made by the water spider. The rat-tailed maggot *Tubifera* is often found in shallow waters with a low oxygen content. This hoverfly larva has an extensible siphon in contact with the surface; the siphon is extended as the water level rises (Fig. 11.15).

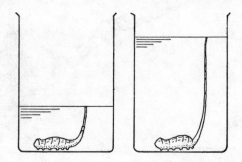

**Fig. 11.15**   Extensible respiratory tube of rat-tailed maggot *Tubifera* (= *Eristalis*)

Pulmonate snails like *Limnaea* and *Planorbis* have lungs which they fill periodically at the surface. The presence of the oxygen carrier haemoglobin in its blood enables *Planorbis* to make better use of the oxygen in its lung.

Small animals like protozoans, whose surface is large in relation to body volume, obtain sufficient dissolved oxygen through their body wall. Larger creatures generally have gills to increase the surface area (Fig. 11.16). Fish and bivalve molluscs have blood gills with fine filaments containing blood vessels. Many insect larvae or nymphs have tracheal gills containing air. These are arranged along the sides of the body as in caddis or mayflies, or at the tip of the abdomen as in

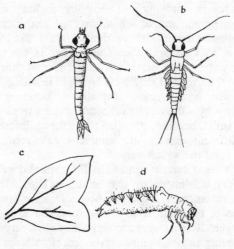

**Fig. 11.16** Adaptations aiding respiration when submerged: (*a*) damsel fly nymph with three caudal gills; (*b*) and (*c*) paired plate-like tracheal gills of mayfly nymph (*d*) paired filamentous gills of caddis larva

damsel flies. Oxygen uptake depends on a good flow of water past the gills, and where the environment does not provide this current the animals have to create their own. To do this, stoneflies and mayflies may jerk or undulate their bodies; caddis flies use a similar action to draw water through their cases. Dragonfly nymphs breathe by means of a 'rectal basket' into which water is drawn and then expelled.

Many rewarding investigations may be made concerning respiratory mechanisms. Structural modifications can be examined and recorded, but often it is the associated behaviour which poses the most interesting problems. How often do air breathers renew their supplies? Does the rate vary with temperature or with activities such as feeding? What affects the rate of respiratory movement in gill breathers? Caddis flies are often difficult to observe, but some will live in short glass or celluloid tubes and others will build cases with pieces of perspex or broken cover-slip so that their movements can then be seen. At the same time this provides a good basis for studies on case building using natural and artificial materials of different shapes, sizes and colour.

### Adaptations assisting reproduction and dispersal

Most water creatures start their independent life as eggs, but *Hydra* may also reproduce asexually by budding, the flatworm *Dendrocoelum* may simply split in two, and the snail *Hydrobia jenkinsi* gives birth to fully developed young.

Egg topics worth studying and recording include: colour, form, number or grouping, any of which may be characteristic for the species; their association with particular places like water surface (*Culex*), stones by a stream (*Simulium*), stones in the steam (caddis larvae), or on water plants (water boatmen); at what time of year they are found and the length of time from laying to hatching; behaviour of the adult when laying eggs; adaptations anchoring eggs under running water – by adhesive, by embedding them in plant tissues, or by filaments which become entangled (stoneflies); differences in shell and yolk at different seasons – the summer eggs of rotifers and cladocerans differ from the winter eggs.

Whilst eggs need certain conditions for normal development and hatching, it is equally important that conditions should suit the young organism which emerges. Thus the beetle *Donacia* lays its eggs on floating leaves of the plant from whose roots the larva later obtains food and oxygen. Because eggs are small they often serve as means of dispersal by being carried in mud on the feet of waterfowl, drifting on detached pieces of water weed, or by some other similar means.

Some animals lack definite larval stages and develop as miniature replicas of the adults. Most aquatic insects, however, have larvae which are strikingly different from the adults and live in different places and eat different foods. Thus dragonfly nymphs hunt in water

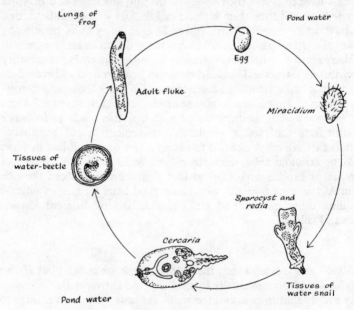

Fig. 11.17  Life cycle of a fluke (*Haplometra*)

whilst the adults hunt in the air; in contrast, the great diving beetle *Dytiscus* which differs greatly from the larva in appearance, nevertheless hunts in much the same way. In general, however, it appears that it is advantageous if larva and adult do not compete for the same resources of food and oxygen.

Parasites may have special larval stages. Thus trematode flatworms spend much of their larval life multiplying asexually in invertebrates (Fig. 11.17). Snail populations are often heavily infected and if species like *Limnaea* are crushed in a watchglass it is sometimes possible to see sporocysts, rediae and cercariae creeping about in the host tissues. Snails suspected of infection can be kept singly and watched daily for the emergence of a cloud of swimming cercariae. Various common pond animals including snails may then be tested to see if the cercariae will encyst in them.

For many water insects, the winged adult is the main means of dispersal to new areas. It has been suggested that winged adults tend to fly upstream to compensate for the aquatic stages being carried downstream by the current. The extent to which animals are dislodged and carried downstream needs fuller investigation.

The absence of a species from a particular habitat does not always mean that it is unable to live there – it may never have had the opportunity. This applies particularly to isolated ponds which have no connection with a stream that might bring in new species. In some cases, the arrival of new kinds of snail has been detected by making thorough periodic surveys of the pond fauna, and it is likely that other new arrivals might be discovered in the same way.

# 12 The Sea and Seashore

The seashore, or littoral zone, is the area between extreme high and low tide marks. Most of its inhabitants have come from the sea, with only a few from the land. Animal life is particularly well represented, and the seashore is a good place to look at animals like sea-squirts and echinoderms which are never found inland. Why are animals like these strictly marine, whilst insects, spiders and others, are rare below high tide level? Such differences in distribution always deserve some thought.

So many species, with varied and unfamiliar adaptations, are crowded on the shore that the beginner may be daunted by the complexity he has to face. The alternation between aquatic and aerial conditions is an added complication contributing to the apparent confusion; however, it all helps to make the shore such a good hunting ground. A day spent there, looking and thinking carefully, will yield many more interesting problems than there is time to solve. Each problem draws attention to one of the countless threads which form the fabric of shore life; the thread we unravel may be small and insignificant, but a deeper understanding can only be reached by patiently assembling such little details.

The following descriptions and suggestions should be regarded only as pointers to the kinds of things which can be done on the shore, and as aids to thinking out your own approach.

## The pattern of shore life

Many littoral species are spread out in very definite patterns; they fall into roughly parallel zones according to their height above low tide level, and are also grouped according to the underlying substratum – rock, loose stones, sand or mud. Detailed records of these patterns are best made by transects – cutting across a variety of substrata at the same level, taking in the widest possible range of levels on a single substratum or (more difficult to interpret) a combination of both.

Shore organisms have little influence on tidal rhythm and exposure which are the dominating habitat factors; thus zonation here mainly reflects different capacities for withstanding exposure, and bears little relationship to plant or animal succession. There are two tide cycles each day, high tide being about fifty minutes later on successive days. The gravitational effects of sun and moon combine fortnightly at the times of new and full moon to produce more extreme *spring tides*; the minimum tides of the intervening periods are known as *neap tides* (Fig. 12.1). Tidal range is most extreme at the equinoxes

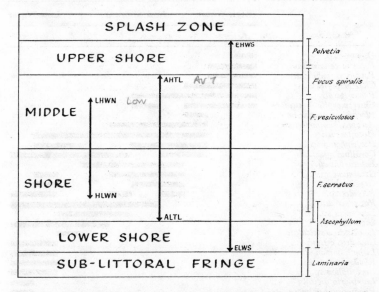

**Fig. 12.1**   Diagram showing neap, average and spring tides in relation to the zonation of common brown seaweeds. LHWN=lowest high water of neap tides; HLWN=highest low water of neap tides; AHTL=average high tide level; ALTL=average low tide level; EHWS=extreme high water of spring tides; ELWS=extreme low water of spring tides

(mid-March and mid-September), so that maximum shore exposure for collecting occurs at low-water of equinoctial spring tides.

Level on the shore is important in determining the amount of drying, heating, or cooling an organism must withstand; it also governs the time available for feeding, and to some extent the violence of wave action. Physical conditions are more extreme near the top of the shore, and this may be verified by taking temperatures, estimating pool salinity and water loss by evaporation at different levels. The way this causes zonation can be demonstrated by moving a large densely colonized stone some distance up the shore, and noting what changes take place. Generally, the number of species per unit area increases towards the lower parts of the shore, reflecting the more equable conditions. In contrast, severe wave action limits the number of species, but extends the effective tidal range for those which can survive.

The zones occupied by different species vary in width and often overlap broadly (Fig. 12.2). Inconspicuous animals and plants are often

Presence or Absence Transect of a Rocky Shore. Bracelet Bay, near Swansea. April 1960

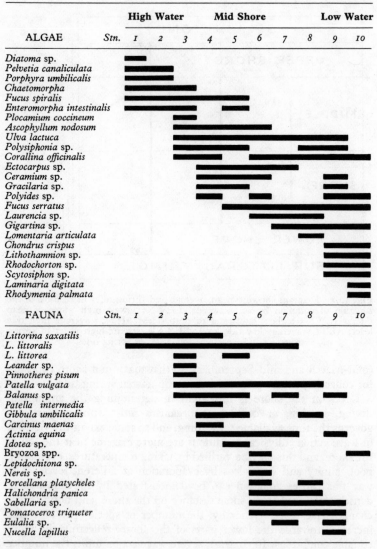

**Fig. 12.2** Zonation of shore animals and plants related to tidal exposure. In a few cases there may be a direct link between an animal and a plant which provides food or shelter.

zoned quite as distinctly as the more obvious species. For practical convenience zonation of conspicuous organisms such as brown algae, lichens or barnacles is used in an arbitrary way to divide the shore when making field notes (Fig. 12.3). Less conspicuous species are then

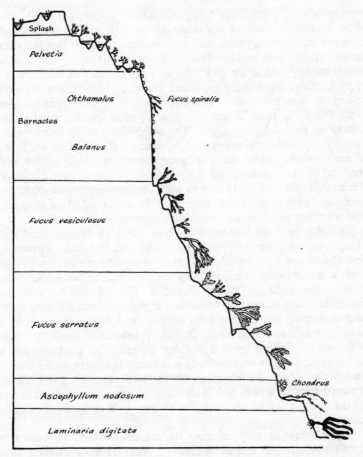

**Fig. 12.3** Profile of rocky shore showing zones of seaweed and barnacle species
In suitable conditions *Ascophyllum* spreads higher up the shore

referred to these zones; thus a species of chiton may occur in the *Laminaria* zone, but not in the *Fucus vesiculosus* zone. An alternative way of dividing the shore is by reference to tide levels; under this classification the same chiton would be said to occur only below low-water of neap tides. The smooth black lichen *Verrucaria*, which is so extensive that it is often mistaken for the natural colour of the rock, is sometimes used to define one of the highest zones on a rocky shore.

## Substratum

Materials are continually being added to or removed from the sea shore by marine currents and wave action. Soft rock is broken and removed more quickly than hard rock, so that the shore line becomes broken up into bays and headlands. The outermost parts of headlands which are exposed to the full force of wave action and currents are often bare and less heavily colonized. However, they shelter the bays and allow loose materials to accumulate and form definite beaches of sand, shingle or mud. Rivers also play a part in beach formation by depositing particles from inland. The composition of the shore therefore depends on several interacting influences – the direction and force of sea currents, the intensity of wave action, the configuration and material of the coastline, and the number and size of river outflows. These influences are fairly easy to work out for a particular stretch of shore, and help to explain why some species are found where they are, and why they may possess particular adaptations.

Basically there are four types of shore – firm rock, sand, mud and shingle, and each has its own characteristic flora and fauna. However, few shores are completely uniform, and on intermediate substrata such as muddy sand or muddy rock the plant and animal associations have a correspondingly mixed character. This relationship with the substratum can be seen by collecting organisms from different areas and comparing the catches from muddy sand and clean sand, or a shingle beach and firm rock. On sandy beaches, animals are best collected by digging up and washing quantities of sand through a sieve; some animals burrow away quickly so it is wise to dig deeply and swiftly. Many of the animals collected will survive better in damp seaweed than in water; it is dangerous to carry glass collecting jars in the hand, particularly on rocky shores, so a rucksack should be used.

The substratum may influence the pattern of shore life both as a base for attachment and as an abrasive. Most littoral organisms have to maintain their position on the shore and many achieve this by some form of attachment adapted to suit the substratum. The holdfasts of large seaweeds like *Fucus* and *Laminaria* are suited to firm rock and are of little use on sand or shingle where these seaweeds are rarely found. Similar considerations apply to barnacles, limpets and chitons.

The abrasive action of sand and shingle when churned up by wave action is strong enough to eliminate many organisms like barnacles and seaweeds, where they come under its influence. Thus large boulders or rocks fringing a sandy beach often have little life on them near sand level, and much more a few feet up (Fig. 12.4). The rock itself may be deeply grooved by abrasion; a comparison of such weathering at different places may be made by fixing blocks of soft, easily scratched material and seeing how they wear. Shingle is even less stable than sand, and the larger pebbles do more damage so that few living things are present except for temporary visitors like gulls, other birds and flies. Most organisms present in sand can burrow to escape the turbulent advancing front of the tide, and because the sand retains water fairly well they can also survive when the tide is out. Shingle does not retain water, and loses most of its organic material through drainage. Microscopic life is often abundant in sand; is this true of shingle?

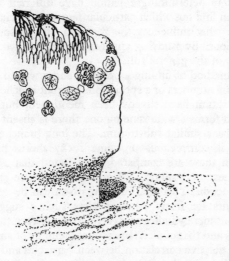

**Fig. 12.4** Diagram showing grooving of rock by sand abrasion, and restriction of algae and limpets to higher parts

The large scale patterning of shore inhabitants into bands and patches is so easy to record that there is a temptation to over-collect data in the form of transects. It is not enough just to recognize and record the existence of zonation; the main problem is to discover its causes in each species. A glance at a transect diagram helps us to pose such questions as 'Why is the barnacle *Chthamalus stellatus* generally found higher up the shore than *Balanus balanoides*?' 'Why are many animals found in the open near low-water mark, but only under rocks when higher up the shore?' It is only by asking and trying to answer such questions that we will begin to understand the reasons for zonation; these may well differ between species. To discover them we must take a much closer look at the details of shore life.

## The search for correlations

When first looking at shore zonation, many people feel disappointed that it is not as clear cut and definite as they had been led to believe. This easily passes into the mistaken view that there is little to be gained from ecological studies on the seashore. However, irregularities of zonation do not indicate chaos, but reveal a complex of interacting processes, some of which we may be fortunate enough to unravel. It is vitally necessary to avoid a superficial and over-simplified approach; irregularities are more an opportunity than an occasion for despair. Thus where an algal zone intrudes deeply into a neighbouring zone this should not be ignored as 'atypical', for it is a useful indication that the factors determining zonation have different values in this area. If we can find out which particular factors have changed we may begin to see what influences the zonation of this species in more 'typical' regions. By studying apparent exceptions we may discover part, at least, of the general rule.

A similar method of finding a problem and a way to its solution is to compare the numbers of a species on different beaches. Differences due to substratum have already been mentioned, but sometimes a species which forms a wide zone on one shore is absent from another although it has a similar substratum. The long branched brown alga *Ascophyllum* is conspicuous on some rocky shores but absent on others. When these are compared, it is found that *Ascophyllum* is absent from exposed shores and grows only where sheltered from severe wave action. This kind of comparison can be used to suggest limiting factors for many other species; such a suggestion is not conclusive, but merely indicates a correlation between the distribution of the species and that of a certain habitat factor or of another species. Thus a clear positive correlation between barnacles and mussels does not necessarily mean that the presence of one directly influences the other; it may well be that some other factor such as the amount of sediment in the water is affecting both barnacles and mussels in the same way, so that they are found in the same place. It is obviously important not to jump to conclusions as soon as a correlation is found, but to look more closely and see if a real relationship is involved and whether it is indirect, as with the barnacles, or direct. In the case of the dog whelk *Nucella*, whose distribution is linked with the presence of barnacles and mussels, the relationship is direct – for careful observation reveals that they form part of its diet. Careful analysis of a correlation, working out its mechanism by theory and observation, is the most difficult part of discovering what limits distribution, and is absolutely essential if a sound and honest conclusion is to be reached.

Close correlations between different species are not rare, so it is worth noting whether one particular animal species is generally associated with another, or with a certain type of plant. Such associations may involve food relationships or shelter. Many small animals, for instance, are specially abundant in the holdfasts of *Laminaria*, which provide good shelter from wave action, predators, and desiccation (Fig. 12.5). Associations of this kind require a great deal of study and much time spent on the shore; yet some of the problems can be discovered and solved in a single day and even if a piece of work only goes as far as establishing that some correlation exists, this is well worth while, particularly if the information is passed on for someone else to follow up later.

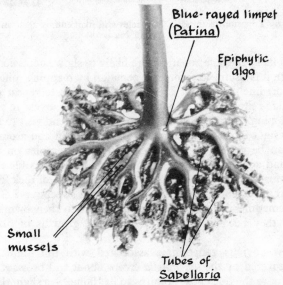

**Fig. 12.5** Holdfast of *Laminaria* giving shelter and attachment for mussels and epiphytic algae besides various crustaceans and tubicolous annelids living in the accumulated silt

An interesting type of correlation occurs when one species is rarely or never found in the presence of another, although the habitat appears quite suitable. Many such examples of negative correlation are encountered in studies of zonation and succession.

## Zonation in miniature

As waves pass over them, large seaweeds brush against the surrounding rocks and may damage smaller organisms; each seaweed has a definite area of sweep within which some otherwise quite common filamentous algae and delicate animals are rarely found. Sometimes this interaction results in a concentrically arranged miniature zonation (Fig. 12.6). As the central seaweed grows and

**Fig. 12.6** Zonation of mussels, barnacles and filamentous algae in the sweep area of *Fucus*

extends its sweep, the zonations are likely to show succession. Hence, although it is useful to record the zonation by mapping, photography or a short line transect, it is even more valuable to repeat the study at the same site after a period of a few months or a year to see how it is developing. If it is impossible to manage a second visit to the coast, some insight into the development of zonation in and around sweep areas can be gained by comparing different examples on the same beach and at the same level. If the central alga is flexible the whole sweep area may be uniformly brushed, although local rock formations may direct currents to produce asymmetrical sweeping. Less flexible algae sometimes fail to sweep the areas close to their own holdfasts, so that they become surrounded by dense growths of filamentous algae.

A less obvious zonation is associated with the limpet *Patella*. When covered by the sea *Patella* creeps about and browses on small algae, but as the tide ebbs it returns to its 'home' – a slight depression in the rock surface. Thus its feeding range extends only a few feet in either direction and the algal flora within this area is often different from that outside it, depending partly on its position on the shore, and partly on the activities of other browsing animals. Limpets affect the distribution of large seaweeds as well as filamentous types by destroying the tiny sporeling stages shortly after they settle on the rock surface; they do not graze the fully grown plants. This can be shown by removing all the limpets from a section of the shore; after a time the algae are found to flourish far more than on other parts of the shore. See also Fig. 12.7.

Rock pools show the clearest and most varied small-scale zonations. Their flora and fauna depends on the size of the pool, its exposure to wave action, and its level on the shore. Pools in the splash zone vary greatly and irregularly in salinity and temperature according to rainfall and sunshine. Large pools may show a compressed zonation of large

**Fig. 12.7** Evidence of mollusc feeding activities. Common periwinkles leave tracks where they browse on microscopic algae and expose the underlying rock. Limpets clear wider areas and do not leave trails. In both cases radula marks may be seen

seaweeds, similar to that on the shore as a whole – thus the pool may be fringed with *Pelvetia*, with *Fucus serratus* and *Laminaria* growing in its depths. Hydroids, and filamentous and encrusting algae show the most conspicuous zonation. Small pools are best recorded by means of a colour photograph accompanied by a rough labelled plan and profile transect. The photograph should always include a ruler or other clear indication of scale, just as dimensions should always be given on hand drawn plans and transects. Zonation above the pool margins should also be recorded, and an attempt made to explain the whole pattern.

## Succession in miniature

Although large scale succession is not found on the shore itself, examples of small-scale succession, which make a fascinating study, can be seen in limited areas. Some must be studied by repeated visits over a long period, but others can be understood quite well by comparing different sites in different stages of development.

New clean surfaces often appear on rocky shores, due to pieces of rock flaking off, and boulders falling from cliffs or being overturned by exceptionally rough seas. These become colonized by young stages of a great variety of animals and plants, many of which are absent from neighbouring rocks. Some species dwindle and die out whilst others become firmly established and some new ones enter the succession. A sequence of bacteria and diatoms, hydroids, a felting of small algae, barnacles, then larger algae, is usual on many parts of the shore. It may be several years before the inhabitants resemble those of older rock surfaces near by. The course of such a succession is often influenced by the season at which the new surface was first exposed, because the young stages of most organisms are available for only a short period in the year, and this differs between species. When one kind of organism has colonized heavily, considerations of space alone make it more difficult for other species to enter.

Because the participants are so small, studies of these successions require at least a hand-lens. If natural colonization sites cannot be found, areas of rock can be scraped clear or test squares can be cemented or bolted on to the rocks. Bolted squares are particularly useful because they can be removed and studied under a binocular microscope in the laboratory.

A peculiar alternating succession sometimes occurs between mussels and barnacles. In some places this is due to predation by the dog-whelk *Nucella*. The whelks, feeding at first on a dense population of barnacles, pay little attention to mussels which settle down in and around the old barnacle husks. These mussels gradually crowd out the barnacles, which are at a disadvantage since they are continually being destroyed by whelks. Eventually there are hardly any barnacles left and the whelks have to choose between starvation and learning to eat mussels. Once they begin to deal successfully with mussels they take little further interest in barnacles, and make large inroads into the mussel population. The empty shells finally break away leaving fresh surfaces for barnacles to colonize. The process continues until there are so few mussels that the whelks switch back to eating barnacles. The record of this dietary oscillation is laid down in stripes

**Fig. 12.8**   Striped Dog Whelk (*Nucella lapillus*) showing white (barnacle diet) and dark mauve-brown (mussel diet) shell banding

on the shell of *Nucella*, which is white on a barnacle diet, but dark mauve-brown when mussels are eaten (Fig. 12.8).

Another type of mussel/barnacle alternation occurs in rather muddy waters. Here uncolonized rock faces do not trap mud, but as soon as barnacles settle in large numbers mud settles in the crevices between them and in the old husks, preventing young barnacles from settling, and possibly choking older ones. This forms a good settling place for mussels, and mud now accumulates rapidly around the byssus threads. An ever thickening layer of mud, byssus threads, and old shells is built up until at last it becomes unstable and breaks away, exposing a fresh surface for barnacle colonization.

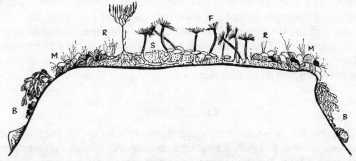

**Fig. 12.9** Diagram illustrating miniature zonation on a flat-topped rock on the lower shore. B = barnacles; M = mussels; R = filamentous alga *Rhodochorton*; F = tube-living *Fabricia*; S = silt

Mud is an important factor in a more complex type of cyclic succession found on flat-topped rocks on the lower shore. Fine filamentous algae colonize the surface, forming a dense mat. If conditions of current, wave action and silting are right, this mat traps mud which becomes inhabited by small animals of various kinds, including mussels and tube-living worms. The algae and mussels diminish as the depth of the mud increases, leaving the tube-living worms as the dominant form. Their innumerable tiny tubes intermingle and bind the mud firmly, bringing about further accumulations until a rough sea breaks the whole mass away and much of the succession begins again. This succession is beautifully mirrored by a concentric zonation with fringing algae and a central area of thick mud (Fig. 12.9).

## Shelter

Many shore organisms need some degree of shelter from harmful influences such as wave action, abrasion, sunlight and predators.

Molluscs and tube-living worms grow or construct their own shelter in the form of shells and burrows; other organisms use sheltered parts of the natural environment. A comparison of life above and below boulders shows this clearly, and the need to replace boulders as found is immediately obvious; nevertheless, a rock could be left upside down and the changes in its population followed as part of a serious study. Rock crevices, empty barnacle shells, and seaweed holdfasts also shelter a great variety of small creatures.

When an organism is found using shelter the important questions are: 'What is it sheltering from?' and 'How does its shelter protect it?' For many shore organisms shelter is important for water-conservation as well as for protection against wave action and pre-dators, and may have particular functions at different times. Thus the moist atmosphere under a rock reduces water loss when the tide is out, and the boulder shields small animals from mechanical damage by the churning of waves and sweeping of seaweeds as the tide returns. Many animals need shelter for only part of the tidal cycle, and roam more widely when the tide is in.

### Life offshore

Marine life is easily accessible down to the upper part of the *Laminaria zone*, but only a boat or aqualung will take us further. The aqualung allows direct observation of sub-littoral creatures and opens up a new field for the biologist, but few have the equipment or train-ing. Trawling and dredging from a boat are rather unsatisfactory because these indirect methods tell us little of how the creatures live.

An easier and probably more rewarding subject to study in offshore waters is the **plankton**, small and microscopic animals and plants which drift at the mercy of the currents. These bizarre and fascinating organisms can be collected in a plankton net towed behind a boat (see Chapter 3, p. 27), but remember that most animal plankton migrates downwards during daylight, and is only abundant near the surface at night. If a boat is not available, quite good collections can be made at night by sweeping a net to and fro in the water a few yards off shore. On some nights numbers of planktonic animals may be captured by simply taking bucketfuls of sea water. Many are luminescent, giving off tiny blue flashes when the water is disturbed, so that surf at the water's edge sparkles brilliantly. These small creatures are very sensitive to temperature change and may not survive even the journey from shore to laboratory unless carried in a vacuum flask or other well-insulated container. A binocular microscope is essential for the proper study of plankton and reveals a new world of strange larval forms and peculiar adaptations for feeding and flotation (Fig. 12.10). Some creatures are permanently planktonic, whilst others are there

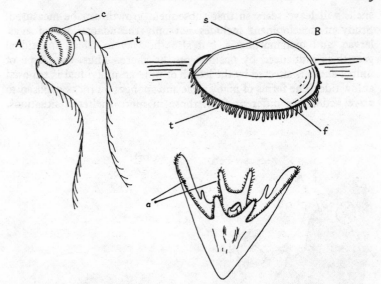

**Fig. 12.10** Feeding and flotation devices: (*a*) sea gooseberry (*Pleurobrachia*); (*b*) by-the-wind-sailor (*Velella*); (*c*) pluteus larva of sea urchin. c=comb-rows for swimming; t=tentacles for capture of prey; s=sail for wind propulsion; f=float; a=ciliated arms for flotation, locomotion and feeding

for only part of their life-histories. Many of the latter eventually settle down on the shore, sometimes undergoing remarkable metamorphoses as in many crustacean larvae, particularly those of barnacles. The planktonic phase is important for both feeding and dispersal, so the shore has close and essential links with the open sea and cannot be regarded as a closed community (Fig. 12.11).

## Other aspects of shore life

Every section of this chapter deals with a different way of looking at shore life, and from each viewpoint it is possible to see a whole host of problems. There are, of course, many other important aspects to consider, such as the adaptations required for feeding and photosynthesis, or the differences between organisms living *in* and living *on* a substratum. In almost any common littoral species, the study of growth and development is worthwhile; size can be assessed by measurements of length, weight or volume, but age is more difficult to estimate – except in some fish where the scales or otoliths have annual growth rings. Shellfish also have growth rings but as these are not necessarily annual, caution is needed. In *Fucus*, age can be estimated from the gas-vesicles which are usually formed only in spring (Fig. 12.12). Notches cut in *Laminaria* stipes or on the edge of mollusc shells will leave scars so that subsequent growth can be measured. Study of development includes not only the adaptations of eggs, larvae, and metamorphosis, but also the way in which individual growth is influenced by position on the shore. Thus, the shape of limpet shells is affected by the length of time an individual is exposed at low tide. The forms of many algae and sponges on rocks exposed to wave action are different from those in more sheltered situations.

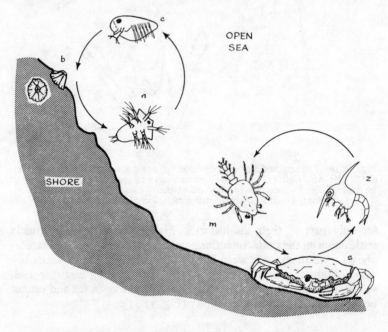

**Fig. 12.11** Links between shore life and plankton of open sea. b = adult barnacle; n and c = nauplius and cypris larvae; s = common shore crab; z and m = zoea and megalopa larvae. Similar pictures could be made for echinoderms, polychaetes and many molluscs

Similarly, the floats of *Ascophyllum* are thicker and more resistant to pressure in plants growing lower on the shore. A different kind of factor affecting growth is seen when parasite infested gastropods grow abnormally large.

Many crustaceans and molluscs are readily marked by cutting notches or applying cellulose paint to the dry shell, and this facilitates experimental study of their behaviour. Marked individuals can be traced at intervals to find how they move between different parts of the shore; in other experiments, animals may be deliberately moved from their usual zone to see if their behaviour is adapted to take them back again. Some animals like *Ligia*, the sea-slater, may need to be followed at night when they are most active and less subject to desiccation and predation. Methods of feeding, especially in sessile forms like barnacles and tube-living worms, can be studied most easily in a

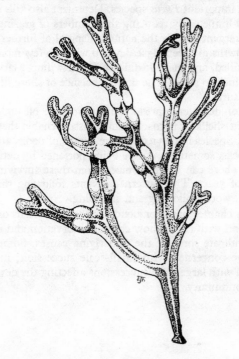

**Fig. 12.12**   Gas vesicles of *Fucus vesiculosus* (see text)

marine aquarium; this should be kept cool, well aerated, and contain no large seaweeds – which decay and pollute the water. Salinity can be maintained by marking the original water level and then 'topping-up' regularly with rain water.

## Pollution

Pollution of the shore with sewage or oil may have dramatic and long lasting effects on the flora and fauna. For proper interpretation, studies of shore pollution need to be based on sound information concerning the previous history of the area. Unusual patterns of distribution may have their origin in events which occurred long before the particular pollution being investigated.

On rocky shores moderate crude-oil pollution on its own has relatively slight effects on shore life because grazing molluscs such as top-shells and limpets gradually remove the oil without noticeably harming themselves. In contrast, widespread use of detergent spraying to remove oil tends to destroy many algae, particularly those found higher up the shore – including the important *Fucus* species. Detergent also kills many animals, especially limpets. By reducing the numbers of grazing molluscs in this way, detergent retards the natural disposal of further oil pollution. Dense growths of green algae develop within a few months of the grazers being killed, and make it difficult for new limpet larvae to find bare rock on which to settle. The normal balance of shore life is therefore slow to return.

The effects of detergent spread well beyond the oil polluted area, but the resistant shells of barnacles and mussels enable these animals to survive well. Species in crevices, in the depths of pools, and beneath overhanging rocks, sometimes escape being poisoned by detergent and the rest of the shore can be recolonized from these surviving pockets over a period of years. The pattern of events following this kind of pollution is well worth following and analysing.

Much of this chapter has been concerned with patterns of zonation on the shore, and with seeing how careful observation and intelligent enquiry may indicate some of the underlying causes. Some patterns were seen to be concerned with small-scale succession; in the next chapter we deal with large scale successions affecting the development of the whole community.

# 13 Maritime Habitats

Maritime habitats are interesting and instructive for study because of the way their vegetation differs from that of habitats further inland. Sand dunes and salt marshes are of particular interest because their condition is rarely static, and in most cases it is possible to see in the one area all the successive stages from the first colonization by pioneer plants, to a fully developed and stable community of mesophytic plants.

## Sand dunes

Sand dunes provide excellent opportunities for studying a modified xerosere (Fig. 2.6, Chapter 2, p. 17), in which the initial xerophytic conditions are associated with the effects of salt spray and unstable sand.

Sand deposited on the shore by the sea and dried by sun and wind is blown inland and deposited to form dunes wherever the wind is slowed down by pioneer plants. As a consequence, the most extensive dune areas in this country are found where the prevailing wind is onshore. In many places, local changes in wind or sea-currents have led to erosion of parts of established dune systems so that the 'textbook' pattern of sand-dune succession described below is to some extent obscured. In exposed areas, continual high winds deposit only the larger sand particles, and this, together with high evaporation rate, means that vegetation is established only with difficulty. A more moderate wind deposits smaller particles which have greater water-holding power, and the evaporation rate is lower. The lighter material deposited may include animal and plant remains from the strand. These often contain, in addition to organic matter, mollusc-shell and crab exoskeleton fragments which significantly increase soil calcium content and thus encourage the activities of root nodule and other bacteria.

There is a well marked succession of plant communities from the shore line, colonized only by salt tolerant species, through the main dunes in which marram grass, *Ammophila arenaria*, is the main stabilizing influence, to the less maritime regions where grassland, heath or scrub may become established.

Whilst plants on the strand line act as wind-breaks and accumulate 'miniature dunes', most have little stabilizing influence because they do not bind the sand effectively. An important exception is sand couch-grass, *Agropyron junceiforme*, which often prepares the way for the entry of marram grass. Sand couch-grass, unlike marram, can stand up to occasional immersion in seawater by spring tides. Its extensive rhizome system binds the sand and it is able to put up new aerial shoots when buried by sand, though it spreads more readily sideways. Sea lyme-grass, *Elymus arenarius*, has similar habits and occasionally forms low foredunes in the same way. The open *Agropyron* community contains few other species of plants, the most frequent being sea sandwort, *Honkenya peploides* (Fig. 9.2(c), p. 137), which has creeping shoots and can itself form miniature dunes.

When the sand level has been raised in this way, conditions are favourable for marram grass. This cannot grow efficiently in very damp conditions but thrives in the drier wind-blown sand. Marram can spread in all directions by means of a branching system of underground roots and shoots. By upward growth it is generally able to keep pace with the deposition of wind-borne sand. The underground systems, both living and dead, hold the sand in a compacted state but cannot fix the air-dry surface soil which is thus liable to be blown away (Fig. 13.1). Dunes formed in this way are called 'mobile dunes' because sand is continually being removed and replaced by fresh

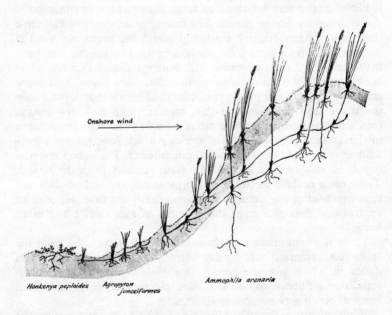

Onshore wind

*Honkenya peploides*    *Agropyron junceiformes*    *Ammophila arenaria*

**Fig. 13.1** Pioneer dune builders. (*a*) *Honkenya peploides* and (*b*) *Agropyron junceiforme* are in the moister, less mobile fore dune and upper shore; (*c*) *Ammophila arenaria* is in less stable, drier, wind-blown sand

sand from the beach. The presence of high mobile dunes reduces wind speed over the lower fore-dunes, and this leads to continued seaward growth of the dune system. On the sheltered side of the dunes the sand is less mobile and a number of plants are able to establish themselves. Many of these are common wind-dispersed inland plants and their distribution depends on the location of seed parents and on suitably damp weather conditions during the period immediately following germination. They include many of the rosette forming plants (dandelions, hawkbits, daisies, cat's ear, sheep's sorrel, storksbill) which are found on chalk downs. Mat plants are also quite common, so that the number of 'dune species' is only a small proportion of the full species list (Fig. 13.3). The abundance of these few species, however, determines the character of large areas. Many help in dune fixation – thus Sand Sedge, *Carex arenaria*, and Sand Fescue, *Festuca rubra* var. *arenaria*, have creeping stolons just below the surface and are able to help fix the surface sand. Others like Sand Spurge, *Euphorbia paralias*, assist in fixation by being deep rooted (Fig. 13.2). Mosses and lichens (e.g. *Cladonia*) also form an important part of the soil cover in this region. Decay of the vegetative parts of all these plants adds humus to the surface soil. The increase in fertility and water-holding power reduces the need for deep root systems and provides suitable conditions for a wider variety of plants; these include many 'winter annual' plants, some perennials and more mosses.

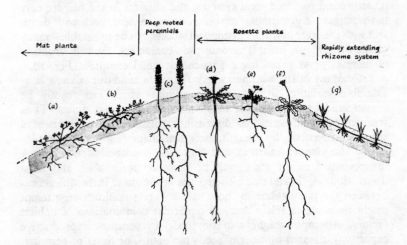

**Fig. 13.2** Dune stabilizers: (*a*) *Medicago lupulina* (perennial); (*b*) *Stellaria media* (annual); (*c*) *Euphorbia paralias* (perennial); (*d*) *Taraxacum laevigatum* (perennial); (*e*) *Teesdalia nudicaulis* (annual); (*f*) *Hypochaeris radicata* (perennial); (*g*) *Carex arenaria* (perennial)

In spite of these additions, surface cover is still incomplete, particularly during the summer drought after the annual plants have died. A strong wind blowing in an unusual direction can remove sand from the usually protected leeward side of the dunes forming hollows or 'blowouts'. These may link up to form valleys or 'slacks' and, where the dune system is based on an impervious substratum, water may collect to form pools or marshes. These moist slacks have their own flora of damp grassland and marsh species. Creeping willow, *Salix repens*, frequently establishes itself in wet slacks and spreads on to neighbouring dunes. Because it can keep pace with accumulating sand blown from mobile dunes, it often gives rise to secondary dunes.

The accumulating plant cover on the landward dunes eventually crowds out the marram grass and gives rise to 'fixed dunes'. The resulting cover may be grassland (used as rabbit warren or golf links, more rarely as pasture). Sometimes a ling (*Calluna*) heath becomes established, or it may be a scrub of bird-dispersed shrubs such as brambles, sea buckthorn, elder, hawthorn, honeysuckle, and privet, together with the wind-dispersed dwarf willow. Such a scrub may eventually give rise to woodland. Even on fixed dunes blow-outs may occur where some circumstance such as burrowing by rabbits has exposed bare sand.

Most sand dunes are suitable for elementary studies. The succession of plant communities may be traced by means of well selected transects (Fig. 13.3), and the interactions of wind, sand and vegetation are clear. Blow-outs provide opportunities for full investigation of underground root and stem systems, and changes in the soil are easy to investigate (e.g. variation of soil moisture content, both with depth and with the state of tide). Anatomical studies can be profitable – many of the dune plants exhibit xeromorphic characters, the leaf structure of marram grass providing a particularly good example (Fig. 10.2, p. 146). Many of the common weeds found on sand dunes show considerable variation in vegetative structure which can readily be investigated on a variety of habitats in and near the dunes. Thus frequency distributions for dimensions (plant height, length of a specified internode, leaf length, etc.), leaf shape (e.g. length/breadth ratio), succulence, hairiness and flowering (number of flowers per inflorescence) might quite well be recorded for mobile and fixed dunes, dune slack, and nearby hedgerow or meadow. If the differences between the populations in these habitats are significant an attempt might be made to relate them to particular combinations of habitat factors. Attempts might also be made to compare reproductive capacity (estimated number of seeds per plant), or times of germination, flowering and fruiting, and relate these to the abundance of the plants in the various habitats.

Compared with meadows or woods, the dunes are poor in animal

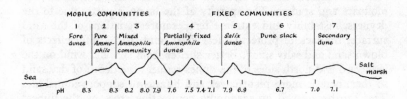

| SPECIES | | COMMUNITIES | | | | | | |
|---|---|---|---|---|---|---|---|---|
| | | 1 | 2 | 3 | 4 | 5 | 6 | 7 |
| *Agropyron junceiforme* | d | X | | | | | | |
| *Elymus arenarius* | d | X | | | | | | |
| *Silene maritima* | d | X | | | | | | |
| *Euphorbia paralias* | d | X | | X | | | | |
| *Honkenya peploides* | d | X | | X | | | | |
| *Carex arenaria* | d | X | | X | X | X | | X |
| *Ammophila arenaria* | d | X | X | X | X | | | X |
| *Arenaria serpyllifolia* | w | | | X | X | | | |
| *Erodium cicutarium* | w | | | X | X | | | |
| *Leontodon leysseri* | w | | | X | X | X | | |
| *Taraxacum officinale* | w | | | X | X | X | | X |
| *Viola tricolor* | w | | | | X | | | |
| *Thymus serpyllum agg.* | w | | | | X | | | |
| *Iris foetidissima* | w | | | | X | | | |
| *Teesdalia nudicaulis* | w | | | | X | | | |
| *Tussilago farfara* | w | | | | X | | | |
| *Arctium minus* | w | | | | X | | | |
| *Clematis vitalba* | w | | | | X | | | |
| *Phleum arenarium* | d | | | | X | X | | |
| *Tortula ruraliformes* | m | | | | X | X | | X |
| *Bellis perennis* | w | | | | X | | | X |
| *Rubus fruticosus agg.* | w | | | | X | | | X |
| *Cardamine hirsuta* | w | | | | X | | | X |
| *Myosotis hispida* | w | | | | X | | | X |
| *Veronica arvensis* | w | | | | X | | | X |
| *Salix repens* | w | | | | | X | X | |
| *Juncus effusus* | h | | | | | | X | |
| *Juncus acutus* | hd | | | | | | X | |
| *Equisetum* sp. | h | | | | | | X | |
| *Mentha aquatica* | h | | | | | | X | |
| *Iris pseudacorus* | h | | | | | | X | |
| *Phragmites communis* | h | | | | | | X | |
| *Potamogeton natans* | h | | | | | | X | |
| *Brachythecium* sp. | m | | | | | | X | |

Key: d = maritime species associated with sand dunes
     w = common plants with a wide distribution
     m = mosses
     h = hydrophytes; plants of marshes and dune slacks

**Fig. 13.3** Transect and presence or absence record—Whiteford Burrows, Gower, S. Wales. (Palmer, N. B. (1960), *Cheltenham G. S. Biology Rep.*, No. 4.)

life. The insects are represented best, with smaller numbers of molluscs and spiders. The sparsity of the fauna is partly due to the dryness, instability and extreme temperature variations of the sand surface, the lack of protective leaf litter, and the harmful effects of blown sand and salty spray. Better shelter from sun and wind on the landward and north sides of the dunes reduces the depth of dry sand so that there is more plant cover and humus to support animal life. This can be verified by examining contrasting areas on the dunes; animals in metre quadrats may be counted by sieving the sand to a depth of 15 or 30 cm and teasing apart marram clumps. Measurements of daily variations in specific microclimates can give further insight into the animals' distribution. For instance temperatures might be taken at various depths in the sand and at various positions in and around marram grass tussocks or under the different coloured layers of moss and lichen.

Many of the animals escape temporarily from the harsh conditions by burrowing down to damper sand, flying or living within tufts of marram grass. The young stages are particularly vulnerable, and it is probably for this reason that insects with elaborate parental or social behaviour are common on dunes. In midsummer, solitary wasps and bees may be seen carrying captured insects and pollen respectively, taking this food to provision their burrows for the larvae. Like the even more abundant ants, they create by their own efforts a limited pocket of favourable conditions within which the larvae can survive. The structure of their burrows can be revealed by careful excavation.

Feeding relationships can best be seen by patiently watching and recording the behaviour of animals in a limited area during the summer. Herbivores are represented by bees, moth larvae and plant bugs; scavengers by ants and the tiny cockroach *Ectobius*, and carnivores by spiders, wasps, robber flies, ground beetles and the tiger beetle *Cicindela hybrida*.

## Estuaries

There is an intermingling of marine and fresh-water conditions wherever a river meets the sea. The flow of silt-laden river water becomes slower and suspended material settles out to form sand banks. The increasingly shallow water over such deposits may be stationary at high tide; this allows deposition of finer material to further raise the level and form mud flats which are exposed for a few hours daily at low tide. These areas may become colonized and develop into salt marsh.

In estuaries near large towns, much of the material settling out is organic matter from sewage or industrial processes. Such material increases turbidity and limits light for water plants, thus reducing the photosynthetic production of oxygen; at the same time, bacteria breaking down this material are further depleting the oxygen supply. Estuarine pollution can thus severely limit the larger forms of life. Digging at low tide in exposed sand or mud often reveals a smelly black incompletely decomposed horizon whose distance from the surface indicates how far oxygen has penetrated. In summer, temperature rising well above that of the open sea may further lower the oxygen content. This is particularly true of pools where the larger and more active animals like sticklebacks are soon affected and die.

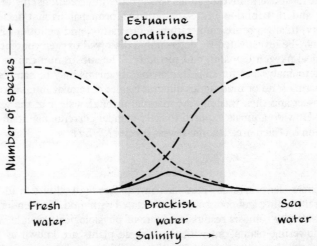

**Fig. 13.4**  Species distribution in an estuarine environment. (-----, Fresh water species; — — — —, Marine species; ———, Estuarine species)

Animals found in estuaries and salt-marsh creeks have to meet many problems, but the most characteristic is the variation in salinity that accompanies each tide cycle. Some animals avoid these changes by living at higher tide levels where there is little dilution of sea water; these are often common seashore animals, and face the normal shore problem of exposure. Others avoid dilution by burrowing into the substratum – there is comparatively little variation in salinity within the pore spaces of sand or mud. Only euryhaline organisms can flourish in the open water. These are either marine and able to tolerate dilution, or fresh-water creatures which can cope with an increase in salt concentration; very few are able to meet the full range of variation from sea to fresh water. Consequently there are fewer species than in either sea or fresh water, but in favourable conditions the biomass may be quite as large; the species present often swarm in immense numbers so that salt marshes and mud flats form rich feeding grounds for wading birds and marine fish which come in with the tide.

Stages in the life history may be affected differently by changes in salinity, some stages being specially resistant or able to escape by migration. This is important in isolated pools, where the salinity often increases as water evaporates in hot weather, sometimes until the pool dries up completely. Thus seasonal comparison of pool fauna with that of drainage channels is likely to yield interesting results. Periodic measurements of salinity will show fluctuations due to evaporation or flooding by fresh water from rain or river on the one hand, and by sea water during exceptionally high tides.

The most useful investigations concern the interrelationships of food webs and distribution in relation to the principal habitat factors – salinity, tidal exposure, substratum, temperature, and pollution. Pollution may be investigated by determining dissolved oxygen content, biochemical oxygen demand, and turbidity. The substratum can be subjected to analysis as for soil. Burrowing organisms may be sampled by excavating sand or mud under quadrat frames at regular intervals along a transect and then washing the material through a sieve at the water's edge. How do animals exposed to river or tidal currents maintain their position? Which area has the fewest species? Why?

## Salt marsh

Pioneer plants invading mud flats have to face high salinity in addition to the instability and poor aeration associated with mud. The tendency to lose water by osmosis results in a state of physiological drought which few flowering plants can withstand; these plants are known as *halophytes* and have special characteristics which overcome the adverse habitat conditions. Many are succulent, having water-storage tissues, and their cell sap has a high osmotic pressure which ensures some water uptake from a strongly saline soil solution. Lack of oxygen is met by the provision of aeration tissues within the plant, and mud instability by the development of long rhizomes or roots.

The principal pioneer is glasswort (*Salicornia* sp.) which is found sparsely distributed in areas uncovered by the tide for the greater part of the day. The plants add to stability and, by helping to check water flow, encourage silt deposition and a raising of soil level. This in turn leads to a general increase in the distribution of the plant.

In the south, glasswort is often replaced as pioneer by rice grass (*Spartina townsendii*).

As plant debris and silt accumulate and raise the level, more plants can establish themselves. Seablite (*Suaeda*), sea aster (*Aster tripolium*), and the shrub sea purslane (*Halimione portulacoides*) are all succulent and can tolerate immersion in salt water; the degree of succulence is related to the time they are exposed to salt water. Sea meadow grass (*Puccinella maritima*) spreads by runners and may form the main constituent of grassy salt marshes.

As the level is raised further, the vegetation becomes more like that of ordinary marsh, but the species are all ones that can survive occasional flooding with salt water. With good drainage, the upper zone may form a rich sea-meadow dominated mainly by red fescue (*Festuca rubra*). Continued rise in level may mean the end of sea-water flooding; rain-water draining through the soil then lowers the salinity so that more and more mesophytic plants can establish themselves.

In most places the picture is less simple than this, so that there are plenty of ecological problems and relationships to investigate. Changes in sea and river current may lead to erosion of established salt marshes so that zonation does not illustrate succession as clearly as in sand dunes. The marshes are broken up by creeks and channels through which tidal waters enter and leave. Old channels are sometimes closed off by marsh formation to leave pools, and new channels are formed. This means that it is rarely possible to study salt marsh zonation by means of a single transect, though sometimes many of the features described can be revealed in a transect across a drainage channel. Separate transects across the different zones should give a clear picture of the plant succession.

Periodic study of a salt marsh will reveal where it is growing and where it is being eroded, and some of the physiographic factors responsible can be investigated. Tidal flooding of a limited area should be mapped to show the time different areas are submerged; this information can then be compared with the vegetation map and determinations of soil salinity. Which animals are associated with the different kinds of vegetation? Are they gaining food or shelter? – and is the shelter from currents, tidal exposure, or predators?

## Shingle banks

Shingle banks are formed of water-worn rounded pebbles which have been thrown up by the sea to form fringing pebble beaches, shingle spits and shingle bars. The fringing beach is in contact with the mainland along its length; the spit is formed where there is a change in direction of the coastline and sea currents deposit shingle on the original line of the shore. A shingle bar is a beach running parallel to the shore line but at some distance from it, often enclosing a freshwater or brackish lagoon (Fig. 13.5).

Shingle tends to be unstable, and this, together with lack of humus, limits the rate at which vegetation can colonize it. Humus is eventually supplied by decay of the few plants which have found a foothold, and by debris cast up by the sea or deposited from the leeward side. Halophytes from the salt-marshes, which often lie behind shingle bars or spits, frequently invade the shingle. Sometimes when these are covered by movement of the shingle they are able to grow up again on what is now the weather side and exert a stabilizing influence.

Where the shingle is mixed with sand many of the species found are those also typical of sand dunes though marram grass, which depends for success on blown sand, is normally absent. Shingle beach communities are open and there are no dominant species which can compare with sand couch grass and marram grass on the dunes.

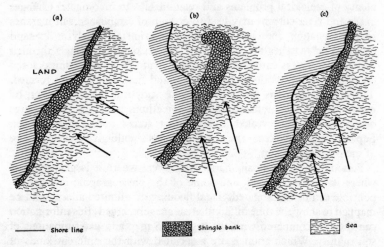

**Fig. 13.5**   Shingle beaches: (*a*) a fringing beach; (*b*) a shingle spit formed by currents along the shore; (*c*) a shingle bar enclosing a lagoon (formed by lowered level of land behind a fringing beach, or by extension of a spit to meet the coast line once more)

Shingle beaches do not afford the same opportunities for studying plant succession as sand dunes or even salt-marshes, though a continued study over a period of years, particularly of shingle spits, can be very rewarding.

Until colonized and stabilized by plants there is little animal life in shingle. The inhabitants of the seaward slopes are active animals and include various terrestrial insects and littoral organisms like sea slaters and sand hoppers which are adapted to both marine and terrestrial life. All these animals are scavengers which migrate up and down the shore with the tide. They avoid immersion, desiccation, and mechanical damage from moving stones by keeping just above water level.

At higher and more stable levels, the plants bring their own associated fauna and there is the possibility of simple food chains becoming woven into food webs.

The lagoon behind a shingle bank often provides excellent opportunities for freshwater or marshland studies. Slapton Ley in Devon is an example of such a lagoon which is regularly studied by groups from the nearby Field Centre.

## Coastal cliffs

Cliffs vary as habitats according to the nature and slope of the rock of which they are composed. Rocks which weather readily bear a richer and more varied flora than harder rocks such as granite. The first colonizers on bare rock are algae, lichens, and mosses. When these decay, the humus collects in crevices and hollows with the mineral products of weathering to provide a soil in which other plants can grow. These are limited by many unfavourable factors, including exposure to sun, wind and salt spray, and the low humus content and shallow depth of the soil. Those plants which do establish themselves are maritime plants showing some degree of salt tolerance and adaptations to conditions of water shortage. Many, particularly at the lower levels, have narrow succulent or leathery leaves with waxy or hairy coverings; they include many of the halophytes found in salt-marshes. Higher up, and on the cliff top, are many drought resistant plants. These include rosette plants, mat plants binding soil by their roots and rhizomes, and cushion plants which have long rootstocks.

Whilst the general impression, particularly on West-Country cliffs exposed to Atlantic gales, may be of stunted grass and wind-blown shrubs or trees, the protection provided by a stone wall, boulder, or even a hummock, may produce a striking contrast in flora. Here, sheltered from blown spray and gales, delicate plants and small shrubs may flourish, some of them flowering for a considerable part of the year. These sheltered areas may provide microhabitats for animals which cannot successfully colonize more exposed areas, though some may range more widely when foraging in mild weather. Such contrasts are well worth investigation by transects or by quadrat sampling.

In many places cliffs are inaccessible, but the cliff line is frequently broken by streams where it is possible to get down to sea level and carry out a vegetation survey without needing to be a skilled rock climber. Inaccessibility has resulted in many sea birds using cliff ledges for nesting; the birds' droppings make an important contribution to the humus available for soil formation. The nests and their associated vegetation can often be observed from other parts of the cliff using binoculars: some of these plants are aliens which have been introduced by the birds themselves.

# 14 Town Ecology

In each of the other habitat chapters, we have attempted to select some feature which appears highly characteristic. In the town, the principal characteristic is the ever-present influence of man's activity, altering normal ecological relationships to create new habitats or down-grade and destroy old ones. The field biologist of the past moved out of town to make his observations – the urban environment was thought to have little to offer. This is unfortunate, for a little thought shows that towns provide good opportunities to meet some of the most important problems in ecology. The roofs and walls of buildings provide surfaces for pioneer plants to colonize (Fig. 14.1), and their ledges and gutters provide similar resources for nesting birds and invertebrates to those found on coastal cliffs. Church towers and high-rise buildings ensure the full range of contrast between extreme exposure and the complete shelter of heated buildings which have for centuries enabled such semi-tropical insects as cockroaches to survive throughout the year. Aquatic habitats vary from blocked gutters on roofs to disused canals and town drainage systems. All these and more provide similar problems to those already considered in earlier chapters. More specifically urban problems include some of the biological consequences of pollution, waste disposal, pesticide use and misuse, and the medical and behavioural difficulties arising from the stresses of town life. In addition to the study of animals and

**Fig. 14.1**  Pioneer plants colonizing brick walls

plants, towns provide opportunities for studying the ecology of man himself.

Few species sharing our towns affect man directly although they may be essential for maintaining the characters of particular habitats; communities of microscopic organisms which maintain soil fertility provide a good example. Some species, including fleas, lice and disease organisms, are directly harmful, whilst others such as rats, mice, and the arthropod pests of stored foodstuffs compete with man for resources. A few are of practical benefit to man, like bees or the freshwater organisms which are used to monitor pollution in town rivers.

As a town grows and changes in industrial, social, and architectural character, so the associated ecological communities also change. This may be the result of such direct intervention as in the management of parks and gardens, or in less easily controlled ways such as the spread of rats and mice in slum areas and mass roosting of starlings on city-centre buildings. It is only through a detailed understanding of the lives of these animals that we can hope to control and limit their numbers. Why are starlings, pigeons and sparrows so much more successful in town than in the country? Answers to this question could have important practical applications. We should equally ask whether *complete* elimination of such birds would bring undesirable side effects.

Social changes affect human ecology more intimately – with the fashion for longer hair, the human headlouse has become increasingly common on males. The spread of central heating has made bedbugs increasingly difficult to eradicate as the warmth enables them to breed throughout the year. The vacuum cleaner has practically eliminated the human flea *Pulex irritans* and led to its replacement on man by the cat flea *Ctenocephalides felis*; the roving habits of cats and their popularity as pets enable this flea to survive mechanized cleaning. Where *do* cats go in their wanderings? Do they have clearly defined territories which they maintain against intruders, or do they mix freely with other cats? Is there a hierarchy of dominance ('pecking-order') among the local cats? What is the function of their night cries? Do they follow regular routes? There have been few studies on the private life of the domestic cat and this is an obvious field for the town ecologist to investigate. Apart from cats, quite a number of animals like birds, bats, and rodents establish nests or roosts on houses. The smaller animals associated with these nests may, as in the case of house moths and the carpet beetle, be of considerable economic importance. A study of factors affecting the choice of nesting site, and of the behaviour and life-history of nest-dwellers will have obvious practical uses.

Trends in the changing patterns of plant and animal communities may be discovered by combining field observations with searches of local records. Museums, natural history societies, and dated maps all provide useful historical background. Observed changes in species found in parks, gardens or waste ground can be related to air or water pollution levels, food availability, weed and insect killers, or the changing use of land.

Parks, churchyards and waste ground form important reservoirs of both animals and plants and generally increase the number of species found in nearby gardens. This can be investigated by quadrat and sweep-net sampling of invertebrates or using standardized methods to count the bird species regularly seen in different districts. The counts should all be made at the same time of day because both birds and arthropods have daily patterns of activity which could otherwise give misleading results. Apart from variation in number, the arthropod *species* found in samples taken in the early morning, at mid-day, in the evening, and at night, may be very different. Railways with their tunnels and embankments, and canal and river banks provide important routes by which animals and plants may penetrate almost to the heart of large cities. Sampling at intervals along such green strips may reveal a gradual drop in species diversity. What factors cause this? Reported sightings of foxes and badgers should be plotted on a map to see if they correlate with such natural features or transport routes. A carefully planned and organized night 'hunt' with many observers may produce a surprising number of new sightings. What are the effects of all-night street lighting on animal movements and behaviour?

## Air pollution

Lower plants often act as useful indicators of pollution and provide interesting problems of dispersal and distribution. It is well known that lichens and mosses are less common in towns and that this is attributed to air pollution. Which are most sensitive, and which can survive town conditions best? By making sets of observations along the lines of roads from windward suburbs through the town centre to downwind country areas, it should be possible to decide how the distribution ties in with other observations (Figs. 14.2 and 14.3). For instance, each record might be coupled with graded estimations of soot on evergreen leaves or painted surfaces using paper tissues and a standardized wiping technique. Alternatively, sets of empty jam jars can be placed for one month at suitable sampling points. The jars are then dried and weighed, and finally, washed, dried, and weighed again. The differences in weight indicate the weight of deposit, which should be expressed in milligrams per month per unit area of jar mouth. Pollution contour maps for a town may be prepared in this way. If wanted, the soot so collected could be extracted with water and organic solvents to determine the nature of substances present. The pH of aqueous extracts may give comparative indications of sulphur dioxide pollution. What differences in both the soot composition and lichen distribution are found between a 'smoke-

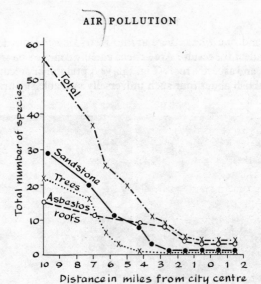

**Fig. 14.2** The richness of lichen flora decreases as the city centre of Newcastle is approached along the transect line shown in Fig. 14.3. (After Gilbert, O. L. 1965, *Ecology and the Industrial Society* (BES Symposium), p. 37, Blackwell

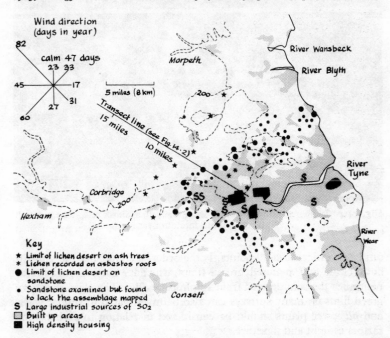

**Fig. 14.3** Lichen distribution in the lower Tyne valley. Compare Fig. 14.2. (After Gilbert, O. L. *op. cit.*, p. 36.)

less zone' and one where there are no restrictions? How do different substrates affect the result? Are lichens equally common on tree-trunks, stone walls, and asbestos roofs (Fig. 14.4)? A preliminary reconnaissance should establish about four such universally available substrates which

**Fig. 14.4**   Lichen colonization of (*a*) stone wall, (*b*) asbestos telephone junction box, (*c*) tomb in churchyard

can be investigated at all points. Is the order of abundance the same in both 'clean' and 'polluted' areas? If not, why not? Can some substrates neutralize the acid derived from sulphur dioxide? The algae that form green films on many surfaces can also be investigated in similar ways, and *all* lower plants should be considered in relation to such habitat factors as light and moisture.

Mosses grow in a great variety of places – not only the more obvious ones such as tops of walls, tree trunks, and on roofs, but at the bases of walls and even in the crevices between paving stones. How are they affected by air pollution, or drainage water? How are they spread? There is evidence that some of the ground-level mosses may regenerate from small fragments carried on shoes. Useful hints of ways to investigate these problems are given in the Nuffield O Level Biology text for Year III.

Higher plants also are affected by air pollution. Conifers are very susceptible; sooty deposits blocking the sunken stomata on their needle-like leaves are not readily washed away by rain. In heavily polluted areas, privet bushes shed more leaves during winter, and many garden plants such as lettuces show stunted growth in city-centre gardens. Careful comparisons between town-centre and suburban gardens should reveal further examples and indicate just how pollution operates to affect particular plants. For a fair assessment, cultural methods and soil conditions in the two areas should correspond; a poor gardener in the suburbs may well produce worse specimens than an enthusiast with a window box in a central office block.

## Parasites and pests

Public health departments may be willing to provide data on which to base surveys of parasites affecting man and animals, and studies on pests of stored products like food and clothing. How are the patterns of infestation changing? How do these relate to control and storage methods? Food is spoiled by arthropods and micro-organisms, some of which (flour beetles, mites) can be cultured in the laboratory so that rates of spread and other characteristics of attack may be determined. How do these features determine the nature of counter-measures? Laboratory cultures of flour mites or similar pests could be kept under different conditions of temperature and humidity to study population change and discover the best storage conditions; the limits of tolerance will vary according to the pest.

Damage to stored products is one example of *biodeterioration*; it is seen again in the economically important damage to timber by fungi such as wet-rot fungus or by wood-boring insects. These infestations can be studied by combining laboratory experiments with field surveys of buildings and fences. What happens to fallen twigs and branches? Is there any link between their breakdown and the attacks on building timber? How effective are the different timber preservatives under laboratory conditions and in normal use?

## Garden ecology

Questionnaires may be helpful in surveys investigating, say, the biology of town gardens. They may be used to question garden owners to establish such facts as: What plants are grown? How does this vary with type of soil, size of garden, aspect, district, occupation and age of owner? What methods of cultivation are used? If garden rubbish is used to improve the soil, is it composted or burned? What pesticides are used, how frequently, and in what doses? Does the method of use correspond with the manufacturer's directions – a question on this point may well reveal the extent of pesticide misuse. Questionnaires may be combined with field investigations to study the effects of pesticides and differing cultivation methods on soil communities and composition, and to determine how the weed flora is related to season, maturity of the garden, and the gardening practices in use.

The habits of garden arthropods pose many problems. What species of insects are associated with particular kinds of flowers, and why? Laboratory and garden experiments using pieces of coloured and grey paper in flower patterns and other shapes, or muslin bags visually masking flowers, may reveal whether the insect's choice is associated with colour, shape, or odour preferences. Is the flower being used for food, or as a place to find a mate? How is pollination achieved and which insects are involved in each case? Relatively inconspicuous but highly scented flowers are often pollinated by night-flying moths.

Much of the life-history and ecology of pests such as 'red spider' mites and aphids can be investigated by mapping at regular intervals the detailed distribution of the species on different parts of plants and in the soil, recording age, sex and reproductive condition. Aphid species are very suitable because their relationships with predators such as hover-fly (*syrphid*) larvae and ladybird (*coccinellid*) larvae are often readily observable, as also are the effects of their ichneumon and braconid parasitoids. Ants often associate with aphids and may obtain 'honey-dew' from them; in some cases they carry aphids to suitable feeding sites. How often are individual aphids visited by ants? What signals pass between the two insects? Can an aphid distinguish between the approach of an ant and the approach of a ladybird larva?

A wider variety of insect and bird visitors can be encouraged by planting shrubs and herbs which have long been naturalised in preference to exotic varieties; a 'wild flower' garden may be even better.

### Insecticides and herbicides

Insecticides may affect soil fertility by decimating populations of soil insects and mites and thus indirectly changing the microbial flora of the soil on which they feed. These effects may be studied by periodic sampling of treated and control garden plots, using the techniques for extracting soil organisms described in Chapters 3 and 7. The consequences of varying the amounts and types of insecticides may also be investigated. Recovery is slow with persistent insecticides; chlorinated hydrocarbons such as DDT may persist in soil for years, whilst organo-phosphate insecticides can disappear in one or two months.

Herbicides are widely used in gardens and parks, or for controlling grass verges and pavement weeds. What herbicides are used locally? Some herbicides disrupt the hormonal control of plants so that the plant grows itself to death. Others block photosynthesis so that the plant starves. Excess herbicides draining into ponds and streams may kill aquatic plants including algae, and some are toxic to fish and crusta-ceans. Root nodule bacteria of leguminous plants are sometimes affected, and high concentrations can cause development defects in animals.

## Energy flow

Direct and questionnaire sampling to study the whole town ecosystem can be used to investigate the total food energy input and the ways in which it is dissipated – to our animal and microbial competitors, in human work energy, in domestic waste, in digestive inefficiency, ex-cretory products and so on. Such a survey might record contents of a representative sample of dustbins which might then be scaled up to give a figure for the town as a whole. How much of this waste was unavoida-ble? Could it be used or reclaimed more effectively? What natural pro-cesses help to remove litter which has not been put in dustbins? How do litter materials such as paper, plastic bags, and matchsticks differ in their susceptibility to biological decay?

## Waste land

In most towns there are patches of waste land and spoil heaps asso-ciated with industrial processes. Some may be new whilst others may be so old that only a search of old records can reveal their origin. What kinds of animals and plants are found there? What is the history of the soil, and its present nature? Where a spoil heap contains only coarse-grained material, it may not have the necessary amount of fine soil at the surface to support seed germination; any fine particles formed by weath-ering tend to be washed away by drainage so that the heap may remain virtually bare for years. In other cases, lack of vegetation may be due to the toxic nature of the waste, as for instance where copper or zinc smelting has taken place. Other urban 'deserts' are the result of heavy machinery compacting the soil so that drainage and aeration are both adversely affected. These possibilities should be borne in mind when try-ing to explain plant and animal distribution.

## Refuse tips

Refuse tips offer varied and changing conditions so that their coloniza-tion by plants and animals makes an interesting study (Fig. 14.5). However, with unstable newly deposited rubbish containing such dangerous materials as rusty cans and broken glass there are real hazards which rightly lead local authorities to exclude unauthorized visitors. It is therefore very important to obtain official permission before attempting to carry out any such field studies.

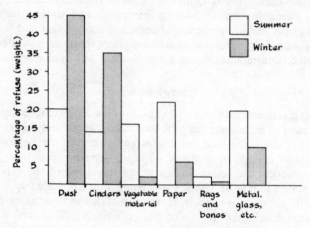

**Fig. 14.5** Seasonal variation in the composition of refuse in a British city (Edinburgh, 1955–56)

Modern methods of controlled tipping mean that refuse deposits are of limited depth and are capped with soil within 24 hours. Though refuse is allowed to settle before the next layer is added, the structure is still sufficiently open for degradable materials to be broken down aerobically to humus. The microbiological activity involved may raise soil temperature significantly so that household plants and animals may survive for some time in an otherwise unfavourable climate. The changes associated with dumping and occasional pollution combine to produce a rapidly altering community. Apart from the normal plants that colonize waste ground, seeds in garden rubbish may give rise to many alien plants which thrive in sheltered parts of the uneven terrain and in the absence of strong competition from native species.

Algae, lichens, and mosses often colonize the soil surface or exposed materials, each substrate being favoured by particular species. Thus green algae such as *Pleurococcus* are found most frequently on damp stones and timber whilst lichens tend to be associated with drier surfaces. Mosses may act as pioneers on compacted soil tracks, paving the way for a succession of higher plants. Uncolonized areas may reflect the presence of toxic material such as substances containing copper.

Where refuse tipping is a means of filling gravel pits or places where minerals have been extracted, there may be fresh-water habitats which become modified as dumping proceeds so that there may be changes worth following in the animals and plants. Where there are decaying plant materials even quite small puddles may be found to contain thriving protozoan populations in which a succession of species becomes dominant in turn. At certain times of year, even the smallest accumulations of water may attract mosquitoes and gnats to lay their eggs. In turn, the resulting larvae and flying adults attract many birds and bats. Other birds are attracted to feed on kitchen refuse or on the soil organisms which such material attracts. The soil organisms themselves may be sampled by hand sorting or by extraction with Tullgren or Baerman funnels (p. 30). Insects or snails can be sampled by 'mark and release' methods and the population estimates may then be correlated with the nature of the rubbish and with birds found feeding in the area.

Garden rubbish heaps are also worth watching; they contain more decaying vegetation than public refuse tips, and occasional bonfires add vegetable ash. *Composting* requires more careful management in which the refuse layers are interspersed by soil, and bacterial activators are used to speed the conversion of plant remains into useful humus. Warmth, moisture, and aeration are important for the bacteria to flourish. Many other organisms occur in compost heaps, and food relationships can be investigated partly by examining gut contents, and partly by observing feeding behaviour under a binocular microscope.

The categories of organism responsible for different stages in decay can be determined by burying dead leaves, grass clippings, or filter paper, enclosed in nylon mesh bags. Very fine mesh will exclude all but micro-organisms, and progressively coarser meshes will admit first small insects, mites and nematodes, with the largest mesh admitting earthworms and the larger insects. The rates and types of decay caused by each category of organism can be determined by examining material from the bags at regular intervals. Losses from materials may be assessed by measuring the areas of leaves or filter paper which have been eroded or by drying and weighing. The method must be decided in advance because the chosen method of assessment must be applied at the start of the experiment on control samples which have not been exposed to decay. Samples of fine sieved compost or rubbish-heap soil may also be investigated using buried cellophane film as described on p. 124.

# 15　Projects

In earlier parts of this book we have tried to explain some of the principles underlying the relationships between plants, animals and their environments, and to suggest many of the ways in which these relationships can be investigated. For a number of habitats we have outlined some of the major features which determine their character; in the course of this we have occasionally posed questions, giving no indication of possible answers. No depth of understanding in ecology can be achieved without making your own study of some limited part of the field, attempting to answer these or similar questions. Thus a detailed study, undertaken as a project, is likely to be more rewarding and satisfying than the more superficial approach of looking at a wide variety of habitats with little time to spare for each.

Projects may concern whole habitats, particular taxonomic groups, individual species or even a single plant like the oak tree as in Chapter 8. Some projects are so time consuming that the best line of approach is to work as a group, pooling results. This may involve the accumulation of results over a long period, as with studies of vegetational change or animal migration, or may involve the division of labour as when a group records the badger population of an area, or when individuals concentrate on their own specialities and assemble their results at the end of short but intensive investigation. Other projects may be designed merely to answer one specific question, and fall well within the capabilities of a single investigator. Such a project may be quite brief, as when watching and recording how a spider builds its web. Thus, however the work is organized, it involves either concentrating on a single species (**autecology**) or looking at a whole habitat (**synecology**).

The examples which follow do not form an exhaustive list, but do at least give an indication of the kinds of project which may reasonably be attempted.

## AUTECOLOGY: Studies centred on single species

Study of the general life cycle is basic; it is surprising how little is known about certain stages in the lives of quite common plants and animals. Excellent examples of what this kind of study may lead to are:

> *Lords and Ladies* by C. T. Prime (1960) (Collins, New Naturalist series)
> *The Ecology and Life History of The Common Frog* by R. M. Savage (1961) (Pitman).

The latter book, in particular, shows how it is possible to discover many new facts even when the subject of study is an organism which has been a laboratory animal studied in both schools and universities for over 100 years.

## Plants

With any organism such a study involves asking and answering a lot of questions. It is convenient to consider plants first under the following sub-headings:

### 1. *Germination*

The main questions are: What are the essential conditions for germination? Do the seeds need to be exposed previously to periods of cold, light or darkness? When do seeds germinate, autumn or spring? What proportion of seeds is viable at this time? For how long do seeds remain viable? (Evidence may come from unexpected plants growing on sites of demolished buildings, or using seeds from discarded herbarium sheets). Is germination epigeal or hypogeal? Is germination affected by depth of sowing?

### 2. *The seedling*

What does the seedling look like – are the first leaves different from later leaves? Early stages of many common plants are imperfectly known, and few descriptions are available. Most seedlings are found in much greater numbers than the mature plants. This implies a high casualty rate – what are the factors causing this, and what are the most favourable conditions for survival? Points to be investigated include both inter- and intra-specific competition with other plants, and the feeding, trampling, or fouling activities of animals (including man).

### 3. *The mature plant*

In what conditions does it flourish most? Points to consider must include shading, altitude, and a whole range of soil factors including water table, pH, soil depth, organic content and mineral content (particularly lime status and salinity). Geographical distribution is important; very often factors which are regarded as limiting, are so only because the plant is near the boundary of its geographical range. Many plants which are regarded as calcifuge or calcicolous in this country display no such tendency in the middle of their range in Southern France or Germany; being at the fringe tends to accentuate the effects of any slightly unfavourable factor.

How efficiently can the plant regenerate after cutting or trampling and what form does this regeneration take? After cutting or grazing, the meristematic regions at the bases of grass leaves are left more or less intact; in broad-leaved plants they are destroyed so that regeneration must come from axillary buds.

Habitat factors are often a cause of 'phenotypic variation' in shape, size and colour of the plant; how does your plant vary in response to shading or changed water status due to flooding or drainage? The answers may often be found by observing the plant in a range of habitats. Many of the variations in size, shape and colouration turn out on investigation to be the result of internal disturbances caused by parasitic animals or fungi. The investigation of host – parasite relationships is of great practical importance in the control of disease and pests where the plants are economically important. The part they play in the web of life has been indicated to some extent for ragwort (Fig. 9.6, p. 142). In what ways does the plant affect the habitat? How much do such effects contribute to ecological succession?

## 4. *Flowering*

What determines the time of flowering? We have seen in Chapter 8 that day length and shading are both important factors – are there any others, and if so, how do they affect the numbers of flowers and hence reproductive capacity? How is the flower pollinated and to what degree is it specialized in this respect? If it is insect-pollinated are the insects also specialized? What size are the pollen grains; are they sculptured, and if so what is the pattern like? What is the nature of the stigma and how does the pollen reach it?

## 5. *Fruiting and seeding*

Flowers give rise to fruits and seeds whose forms are clearly related to their modes of dispersal, and where these are inefficient, survival of the species may depend on a high level of fecundity. Thus for any plant we might ask: What are the fruits and seeds like, and how are they dispersed? How effective is this dispersal – what proportion of seeds reach a situation where there is some chance of germination and survival? Measures of dispersal can be made using sticky traps (p. 24) to collect seeds or fruits at various distances from the parent plant, or by investigating the frequency and effectiveness of contact with animal dispersal agents. Does fruiting reduce the probability of survival to the following year? Does grazing which removes flowering and fruiting heads promote perennation?

In studying the plant, vague generalizations are of limited value, so wherever possible one should make quantitative records of dimensions or numbers, and record the kinds of variation seen within or between populations. These might include mortality, pathogenicity of fungal infections, and various aspects of competition including reproductive capacity. The latter involves counting the number of seeds produced per plant and scoring the results in number-classes to obtain a frequency distribution from which to determine the mean and standard deviation of the population (see Chapter 5). The distributions found in two or more populations may then be investigated for significant differences by calculating the Standard Error of the Mean (p. 76f). If there are significant differences, it may be possible to relate them to habitat factors such as shade or soil pH.

## Animals

Though animals differ from plants, many of the basic problems are similar, particularly those referring to the life-cycle. Thus we may ask: How, when, and where does the animal begin its life? Does it start as an egg or is it born? If an egg, how large is the food store? Is there a larval stage? At what time or season of year does the life-cycle commence, and in what conditions? What are its food preferences, and do these differ between juvenile and adult stages?

We need to consider such features as:

Death rates at various ages, and their causes.

Details of development and growth (including, for arthropods, precise description of ecdysis).

Methods of asexual multiplication, e.g. larval reproduction (as in flukes), budding (as in *Hydra*), parthenogenesis (as in aphids).

Time scale of the life cycle, with information about stages sensitive to drought or cold, or the avoidance of unfavourable periods by encystment, diapause or hibernation. Where are resting stages found?

Seasonal succession of feeding habit, and of gregarious and solitary or territorial behaviour in relation to feeding and reproducing. Are larvae more gregarious than adults? Does either stage feed on more than one species? If so, is food taken from more than one trophic level? Is the food fresh (living) or dead?

Dependent organisms – the nature of predator/prey or parasite/host relationships, as well as the effects an animal may have on the vegetation through grazing, stripping bark, trampling or use of regular latrine areas. What information can be obtained from 'droppings' such as faeces, owl pellets, and debris beneath roosts?

Population studies, animal behaviour, and selected groups are considered in more detail below.

## Population studies

Life tables, on which population trends may be predicted, can often be constructed after collecting information about fecundity and the death rates at different ages. As an illustration of the method, a wood-louse colony from under an isolated stone slab or log could be investigated in early summer by taking a total census and scoring numbers in size groups based on body length. When these results are plotted as a histogram, the distribution will be seen to have two distinct peaks representing the two oldest age groups, so that good estimates can be made of the number in each. A study of the life history may show that this species of woodlouse only reproduces in its third year of life, and does not survive into the fourth year. A closer look at the colony will show that many of the larger (third year) females have a number of tiny babies clinging to them. A count of the numbers on each of a number of females provides an estimate of the average fecundity, and this figure multiplied by the total number of third year females completes the table to give the number in each age group of the current year. The *survival rate* from each year is the proportion of the age group surviving to the following year, so assuming a stable population, survival rates may be estimated as follows:

survival from year 3, $S_3 = 0$ (no survivors to year 4)

$$\text{survival from year 2, } S_2 = \frac{\text{number of year 3 individuals}}{\text{number of year 2 individuals}}$$

These figures have been obtained from a single census and can only be used for forward prediction if the ratio of population change is known. This is obtained by carrying out a census the following year; the ratio of census figures in the two years is used to correct the estimated survival rate. If the total population increases from year to year, then the age group survival rates must be increased proportionally.

Hypothetical example:

A first census gives 250 woodlice in two size groups, there being 150 of the smaller individuals. A count of babies suggests an average of 15 per adult female. Assuming that half the largest-size group is female, this gives a youngest-age group total of 750 and a total population of 1000. Survival rates calculated from this are:

$$S_1 = \frac{150}{750} = \frac{1}{5}; \ S_2 = \frac{100}{150} = \frac{2}{3}; \ S_3 = 0.$$

A brief second census in Year 2 gives 225 individuals in larger size groups, so ratio of population change = 250:225 = 10:9. This makes possible the preparation of a life-table with forward predictions:

| Year | 1 | 2 | 3 | 4 etc. .... |
|---|---|---|---|---|
| Age-group 1 | 750 | 675 | 612 | |
| Age-group 2 | 150 | 135 | 117 | |
| Age-group 3 | 100 | 90 | 81 | |
| Total present | 1000 | 900 | 810 | 729 ... |

Such studies require a full knowledge of the animal's life history. Predictions are only valid if the rates of population loss and gain remain unaltered; prolonged periods of dry or wet weather or other abnormal conditions can drastically affect the composition and stability of a population, producing results very different from those forecast.

### Animal behaviour

The behaviour of an animal often reveals significant features of its ecology. It is important to make the initial study in the field, or under a reasonably full simulation of natural conditions, because confinement in the laboratory usually suppresses or distorts some aspects of behaviour. Detailed records of behaviour in the field not only give clues to its ecological functions but also help in designing laboratory experiments to discover which stimuli trigger and direct th various activity patterns.

The use of movements and postures as social signals is best seen in fishes and birds. Observations on ducks and geese at local lakes or reservoirs will provide a wealth of data, and by recording the detailed sequences of activities it is possible to discover the 'meaning' of some of the signals. This is done by listing how frequently certain kinds of action, such as attacking, escaping, or mating, follow a particular signal: thus a posture often followed by attack may reasonably be designated a threat posture, particularly if it also induces retreat in an opponent. Fish behaviour is more easily observed in aquaria than in the field; for example the territorial, courtship and parental behaviour of male sticklebacks might be investigated experimentally by varying space and the number of other fish, and by using models to determine what features or movements of a partner trigger off particular types of reaction.

Precise objective description is essential; it is all too easy to be vague and subjective about animal behaviour. Terms like 'agitation', 'sleepiness', 'interest', or 'anger' are inadequate. Each posture or movement should be carefully defined in terms of the motions or relative positions of the animal and its parts. Records of locomotion should specify its type, direction, speed and frequency.

Long apparently complex sequences of social or feeding behaviour can be summarized in a flow diagram showing how often each particular unit of behaviour is preceded and followed by each of the other units. Such diagrams often make it easier to see the overall pattern and to identify the key problems. Figure 15.1 shows how the diagram is constructed.

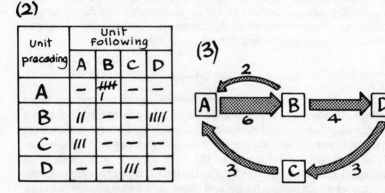

**Fig. 15.1**   Three stages in constructing a flow-diagram of behaviour. (1) Acts and postures are defined, given code letters, and recorded in the order in which they happen. (2) From this sequence a table is drawn up scoring how frequently any particular unit is followed by any other. Thus at the beginning of the sequence B follows A, D follows B, and so on. (3) The table is then used to build up a flow-diagram in which thickness of arrow can indicate frequency of connection between behaviour units. Normally some hundreds of observations would be required to reveal significant patterns.

Behaviour also plays a vital part in the relationship between a predator and its prey, several aspects of which can be studied easily. Thus birds feeding on a lawn might be familiarized for a few days with artificial prey objects. A mixture of plain flour and cooking lard (about 5 : 2 by weight) will produce acceptable bait which is easily coloured by the addition of food dyes and is soft enough to be moulded into different experimental shapes – a cake-icing nozzle will form excellent 'worms' or 'caterpillars'. By varying colour and shape, experiments can be carried out to investigate the effects of resemblance or contrast on predatory behaviour. Such experiments may concern disruptive patterning, resemblance to background colouration, mimicry of inedible or distasteful objects or insects, shape, colour, size, patches of brilliant colour such as eyespots, inedible components, or rareness in the midst of an equally acceptable, commoner, and differently coloured prey. How does predatory success vary with clumped, random, regular, close or

sparse spacing of the prey objects ? What searching patterns do predators employ and do these vary according to the availability and nature of
the prey ? What do the findings suggest about the ecology of prey animals and the limitations of predatory techniques ?

An important function of behaviour is to keep animals in the more
favourable parts of their habitat. The stimuli which determine an animal
species' characteristic pattern of distribution can for most invertebrates be identified in the laboratory using simple pieces of apparatus
which allow the animal to make a differential response when faced with
alternative conditions. For example, the animals might be strewn
evenly along a tube or trough graded in temperature along its length.
At intervals the changing distribution can be noted, taking care that
extraneous stimuli, such as light, are not used as cues for aggregation.
Most species will have one or more temperatures at which they tend to
accumulate.

A more versatile piece of apparatus is the choice-chamber, constructed
from Petri-dishes. Contrasting conditions of humidity, moisture, odour,
texture of substratum etc. can be set up in the lower dish, then animals
introduced through a hole in the lid. The simplest version is floored with
filter paper on which the animals can crawl. Areas of the filter paper can
be separated by using a hot needle to impregnate a thin line of wax;
then by adding drops of a different solution to each, odour, taste and
moisture preferences may be revealed. Alternatively the lower dish may
be divided into two compartments (Fig. 15.2) which are filled with solutions controlling the relative humidity of the air above (various concentrations of caustic potash will do this but must be used with great
caution). Animals are released into an upper chamber floored with fine
mesh voile, or perforated zinc, and will aggregate at the preferred humidity. By varying the intensity of light on the two sides, preferences for
humidity and light combinations can also be investigated.

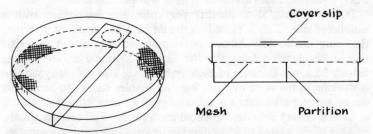

Cover slip

Mesh          Partition

**Fig. 15.2** Choice chamber. The upper chamber has a mesh floor attached,
and an opening at one edge for easy introduction and removal of animals. This
hole can be closed by means of a cover slip. If the chamber is constructed from
a plastic Petri dish, the mesh may be attached using perspex cement, and the
hole bored with a heated cork borer. The lower chamber is divided by a partition
which just touches the mesh. The chamber is normally illuminated from above
and shielded from side light.

Responses to odours can be studied by placing the animals in the stem of a Y-shaped tube (olfactometer); equal currents of air pass down each arm to the stem, but one current has passed over some odorous substance. The choice of arms can be recorded as it occurs, or a collecting trap can be included on each arm so that the counts may be made at the end of the experiment. This apparatus can easily be adapted for use with aquatic creatures.

It is helpful to watch the behaviour of the animal as it makes the choice in any of the methods just described. Thus the location of the relevant sense organs may be revealed by special movements of the structures bearing them, or the mechanisms of aggregation may be found to include obvious directional responses or more subtle effects influencing speed of movement or rate of turning.

## Bats

There are about a dozen species of bats in Britain. The best place to find and study them is in the caves, disused mines, roof spaces and hollow trees where they roost by day and during their winter hibernation (Fig. 15.3).

A project easily carried out by one person is to observe a particular roost throughout the year and note the changes in population, frequency of waking (especially during hibernation), favoured roosting positions and postures, food preferences (from study of insect débris and droppings), and time and place of feeding flights in relation to light intensity, temperature and availability of prey. The bats should not be disturbed by handling when making such a study. To study breeding habits, however, handling *is* involved because it is necessary to weigh each bat and its offspring and record its sexual condition – for instance, 'female, lactating, carrying one offspring'. Parasites, such as ticks, nycteribiid flies and fleas should also be recorded.

It is very helpful to identify particular bats by ringing with a numbered ring. This is placed on the forearm but must not puncture the wing membrane. Advice on ringing techniques and materials should be obtained from the Bat Group, Mammal Society of the British Isles, c/o Institute of Biology, who will also say if anyone else is working in the same district. Ringing enables studies to be carried out on a larger scale over a number of years by organized groups such as natural history societies or school sixth forms. Full records should be kept of each bat ringed preferably by card index using a separate card for each individual. The journeys of bats from roost to roost may be traced by visiting all known roosts in a certain area several times a year, and ringing all bats found. Roosts used for breeding usually differ from the hibernation sites, so that a seasonal pattern of movements emerges. It must be remembered that some journeys may be a consequence of handling by the bat-ringers, so it is necessary to be cautious in interpreting results. Studies of the way in which bats are able to find their way across country to a new roost would be very interesting. Since they are very delicate creatures, they must be handled with great care. It is inadvisable to release them in the open during the daytime because they are likely to be attacked by birds.

**Fig. 15.3** (*left*)   Lesser horseshoe bats and a cave spider in a Gloucestershire stone mine. (Photograph by R. D. Ransome)
**Fig. 15.4** (*right*)   Whiskered bat; note the ring, which was obtained from the Mammal Society and bears a London Zoo number. (Photograph by M. D. Hayward)

### Small-mammals

Voles, shrews and mice are the only mammals small enough to be caught regularly in a Longworth small-mammal live trap. Since they are shy and retiring creatures, two methods of study are open – to keep them in captivity or to sample populations in the field.

In captivity it is easy to watch their behaviour. Fighting, mating, nest-building and care of the young can be seen very well if the animals are kept in a large dry aquarium, floored with an inch or two of soil and rocks. It is best to avoid describing behaviour in general terms like 'the animals fought', but to note precisely the sequence of acts carried out, such as: 'sniff opponent, turn away, face opponent, rear up on hind legs, jump forwards, bite.' The postures seen may be sketched. Sometimes one animal appears to dominate the others and it is worth noting if this is in terms of combat, food, mates or territory.

What functions does this behaviour have under field conditions? Has it any relation to breeding state?

Studies in the wild usually involve trapping animals alive. The traps should contain nesting material and food such as oats to keep the animals in good condition, especially in winter. The nightly range of exploration can be found by trapping and marking the animals and then re-trapping over a period of weeks, with the traps set over a wide area. Homing ability may also be investigated – the animal is caught, then released some distance away from its normal range, where traps are set to record its return. The capture-recapture method (see Chapter 4, p. 67) can give an idea of the total population, and at the same time provide information about breeding condition and the age structure of the population in different seasons. The animals may be marked by clipping small pieces from the ears, or not more than the last joint from a toe, using a local anaesthetic. If done carefully these methods cause far less injury than ringing. Details of marking methods are given in the *Handbook of British Mammals* (Ed. H. N. Southern, Blackwell 1963).

The system of runs and nests can be traced by digging (with the land-owner's permission), and investigations might include comparison of summer and winter nests, and looking for any distinctions into breeding, sleeping and food hoarding nests. By putting out dishes with fruits and seeds of various kinds, and shielding against birds with a little wire netting, food preferences are shown by the types of seeds remaining uneaten. These apparent preferences and dislikes may be tested with captive animals.

The grazing activities of voles are very important in establishing and maintaining the pattern of tussocks and intervening spaces commonly found in rough grassland. Voles tend to move in the interspaces whilst many invertebrates live and hibernate in the tussocks.

### Badgers

The badger is common in most parts of Britain especially in woods, and with patience is quite easy to study. The interconnected burrows, 9–14 in. in diameter, form a *sett*, from which the badgers emerge just after sunset. If a dark-clothed observer approaches quietly so that the wind blows his scent away from the sett, and then sits against a tree to hide his silhouette, the cautious badgers will usually emerge without suspecting. In spring and summer, chasing, play and mating can be seen for a while before they move off to feed. Full notes of the behaviour should be made on the same night, and times taken with a luminous watch. Observations throughout the year will show first appearance of cubs, months when mating occurs, when the sett is extended, and the arrival of strange badgers from other setts.

Flashlight photographs are useful for recording behaviour. One method is to set the camera on a tripod at 'bulb' or 'time', open the shutter, set off a flash bulb, then close the shutter again. With most bulbs this represents a shutter speed of 1/25th s, so that movement by the badgers causes blurring. Alternatively a flash gun can be used for synchronization, allowing shutter speeds of 1/100th s or higher so that a tripod is unnecessary. Strips of bright tin nailed on to trees around the sett familiarize the badgers with the shiny flash reflector. It is best to focus beforehand on the spot where the animals are expected to be, since it is very difficult to estimate distances in the dark (Fig. 15.5).

**Fig. 15.5**  Badgers following the photographer's trail. (Photograph by N. B. Palmer)

A number of observers, by watching all setts in a wood, can determine its population. Once the ratio of number of badgers to number of holes is known for a limited area, counts of holes over wider districts can give an approximate idea of the total badger population (Fig. 15.6). Badgers leave many signs of their nightly activities. Areas important to them can be found by mapping their path systems, which may lead to other setts, feeding grounds or water (Fig. 15.7). These are also shown well by tracks in the snow. Earth scrapes indicate where they have been digging for food. The dung pits by each sett yield material which can be washed and sieved for beetle elytra, small-mammal bones, vegetable fibre, etc. and examined microscopically for earthworm chaetae or the earthworm parasite *Monocystis*. In this way, and by dissecting badgers found dead, the diet can be followed seasonally. Outside the burrows may be piles of fresh or soiled bedding – usually dry grass and leaves. What determines renewal of bedding or material used? The bones and parasites thrown out with the bedding are also of interest. A plan of the sett, showing holes,

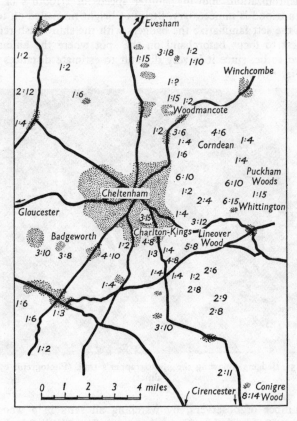

**Fig. 15.6** Map showing distribution of the badger around Cheltenham. Built up areas = shaded; main roads = thick lines; first figure = no. of setts, second figure = no. of badgers—e.g. 3:18 = 3 setts, 18 badgers. [From Humphries, D. A. (1958). 'Badgers in the Cheltenham area.' *School Science Review*. No. 139.]

dung pits and paths, usually reveals effects on the surrounding vegetation. Nettles and elders are typically present. Why?

The Mammal Society of the British Isles has conducted a National Badger Survey and has a record of members interested in badgers (and some other mammals) who can be useful local sources of advice and information.

**Fig. 15.7** Badger path, distinguished from ordinary footpath by the way it passes under fences

### Other wild mammals

Other mammals which might profitably be studied include the mole – found in practically every land habitat except very acid peat, stoats, weasels, squirrels, hedgehogs, rabbits, hares, and foxes. Molehills should be plotted and recorded; is there any relationship between their numbers and earthworm population estimates? Is there any recognizable pattern governing the times and intervals at which molehills appear in different areas?

Rabbits are said to be increasing in number and reestablishing themselves in areas where they have rarely been seen since the myxomatosis epidemics of the middle 1950s. Are their numbers stable, or increasing? What size communities do they live in? Is there any truth in the assertion that many rabbits no longer live in burrows? To what extent does this idea originate from the mistaken identification of hares as rabbits? Is myxomatosis still prevalent? If so, what proportion of victims survives?

### Domestic animals

Animals do not lose ecological interest simply because they have been domesticated. Much can be learned by watching their behaviour. Grazing animals like sheep and cows do not move at random in their field. Things to record include periodicity of feeding, selection of food, tendency to group or to split into sub-groups, whether particular individuals determine the activities of the others, and whether there is dominance or territorial behaviour. The mother–young relationship is also of interest; how does this affect herd grouping, how does the female react to strange offspring and how do the young recognize their own mother?

In towns, the activities of dogs and cats can be followed. Even aquarium fishes yield plenty of data. A good place to see cats is where kitchens of institutions such as hospitals throw out scraps. Large numbers of strays come for food and show various aspects of social behaviour. Useful observations may also be made if you have access to an animal shelter or kennels. In any particular species what are the features we call 'tameness'? How does an animal become tame and what functions does this behaviour serve?

### Birds

In spite of the popularity of ornithology, there is no lack of problems. General recording of species seen does not lead far; what is required is investigation of specific features of bird life. Field-glasses and a notebook are most essential.

Methods of estimating numbers vary according to species and habitat. In woodland an observer may walk along a number of line transects recording all birds seen or heard. If the transects are parallel and sufficiently close, the total population of the wood can be estimated. Care must be taken in comparing results between species because song or colour make some more conspicuous than others. The tendency to sing or feed varies during the day so that a given species may appear to be more abundant at certain times. Most birds sing at dawn. Their ability to hide more effectively in certain habitats must also be considered. It is best to estimate seabirds' numbers in the breeding season when they gather at limited sites.

The behaviour of birds has received much attention. Feeding, territorial and breeding behaviour are of great interest. Close observation of individual birds is better than gathering scraps of information from several sources. Studies may well centre around a nest. How does pair formation occur? How is the partner distinguished from a potential rival? How is the nest site selected and what kinds of interaction occur with neighbouring pairs? What functions do the two parents serve during the nest-building, brooding and feeding phases? What kinds of food do the nestlings receive and how often? How does each gain a fair share? How do the young birds become independent and learn to feed themselves? What functions do bird calls and coloration serve? What interspecific effects can be seen? Some of the answers may be obtained more easily if nest boxes (Fig. 15.8) are provided. According to where they are sited, many different species may use these boxes. The opportunities which they provide for inspecting nests and their contents may also facilitate the collection of ectoparasites (fleas and lice), and commensals like scavenging moth larvae and beetles. It is obviously important to time and limit the number and duration of nest-boxes inspections in such a way as to avoid making parent birds desert the nest or fledglings leave before they are ready. Bird tables and bird baths also provide opportunities for studying the variations in behaviour between species, and between individuals of the same species.

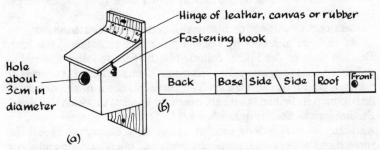

**Fig. 15.8**  Nest box. (*a*) Complete; (*b*) how to cut a 15 cm (6") plank to make it

Feeding habits of birds of prey may be deduced from the contents of regurgitated pellets. The latter may be gently teased apart and such studies are particularly suitable in the case of owls which have fixed roosting places beneath which pellets containing identifiable remains may regularly be found (see Fig. 9.5(b)). *The Handbook of British Mammals*, p. 144, has a key to skulls and lower jaws of small mammals found in owl pellets.

Ringing is useful both in relation to national research schemes, and for intensive local studies. However, it is essential to receive proper training from qualified bird ringers.

### Parasites

The techniques of study vary with the group chosen. In plant parasites like dodder, mistletoe or fungi, it is worth looking at dispersal mechanisms and how the young plant establishes contact with the host and begins to take food. Dodder may transfer from one host to neighbouring plants; is the primary host always the same? Field observations can be linked with sectioning of tissues and microscopical examination in the laboratory.

Among the many kinds of animal parasite, we may select flukes and fleas for special consideration. Flukes are interesting because of the way their larvae multiply asexually in molluscs. This can be studied by crushing the tissues of a water snail in a watch-glass and examining them under a microscope (see Chapter 11, p. 171). The life cycle can be followed in the laboratory, starting either with larvae from pond snails, or with eggs from adult flukes kept in physiological saline such as Ringer's solution. Various possible hosts can be placed in filtered pondwater with miracidium or cercaria larvae, the hosts being dissected later to look for infestation. Adult flukes, usually only a few millimetres long, are best obtained by examining the gut contents of vertebrates, or the body surface and gills of fishes and amphibians. For identification *The Trematoda* by Ben Dawes (1946, Cambridge) is useful.

Fleas occur on birds or mammals. Both larvae and adults can be found in large numbers by searching the nest or home of the host. Disused nests or dead hosts collected in the field should be placed in polythene bags to prevent fleas escaping. Identification with records of numbers and host can be useful, for little is known of the details of distribution in Britain; there are over fifty species, to which the Royal Entomological Society publish a key*. Studies on limited areas, by searching hosts and nests without undue disturbance, can reveal the correlations of fleas with particular hosts and microhabitats and show seasonal changes in numbers and distribution. For instance, the small-mammals of an orchard could be regularly trapped and de-infested, or a house-martin colony could be examined at intervals. Such studies naturally lead to questions like: How does the flea spend the winter? How does it reach the correct host? What are the host's reactions to the parasites?

## SYNECOLOGY: Studies centred on the community

Since suggestions for the study of many common habitats have already been made in earlier chapters, we list here more limited problems.

Changes in the numbers and kinds of plants and animals are best followed through for at least one year. The following are suggested as being worth particular attention:

### 1. Road verges

Selected verges of country roads often show interesting patterns of distribution (Fig. 4.8). What are the pattern-determining factors? Are they constant throughout the year? What differences are found between North and South facing verges? How is roadside growth con-

* Smit, F. G. A. M. (1957). *Handbooks for the Identification of British Insects*, Vol. I (*16*) *Siphonaptera*, London, R. ent. Soc.

trolled? Is it by regular cutting, use of herbicides, or a combination of both? Which animals and plants are favoured by this treatment? What differences are there from untreated areas? How are neighbouring hedgerows and fields affected by the herbicides?

## 2. *Walls, pavements, industrial waste heaps* (*china clay, coal mines etc.*) *and screes*

All these provide opportunities to study colonization by pioneer plants and trace the natural succession. As in the study of sand-dunes and salt-marsh, it is often possible to see most of the stages of succession in the one area, and local enquiry may help to establish times at which different surfaces were first exposed. Walls and pavements are colonized in much the same way as natural rock surfaces (Fig. 14.1). The primary colonizers are often lichens and mosses which pave the way for flowering plants by contributing to the breakdown of the surface and providing organic matter which retains moisture. Which part of a wall is attacked first – stone, brick, or cement? What is the texture of the material in waste tips (see Chapter 7)? Is drainage excessive, or impeded? What are the species involved in the succession? Is there any evidence as to the time scale involved? Does colonization occur more or less readily on shaded surfaces; North-facing, or under trees? What prevents the colonization of bare surfaces? Are seedlings or sporeling stages specially vulnerable and can transplanted mature plants survive? Are toxic materials present?

## 3. *Newly dug soil*

In newly dug soil, the reservoir of buried seeds already present may be far more important than colonizers from outside. Interesting contrasts may be made between the appearance of weeds on soil that is dug or ploughed annually, and those that appear in soil that has been dug for the first time in many years.

## 4. *Small freshwater habitats*

Small pools may collect in many places such as tree-holes or partly blocked gutters. The organic debris which collects in such places provides a feeding resource for a variety of organisms including copepods, rotifers, protozoa, and at least three species of mosquito. How do such habitats vary through the year? The small volume of water involved must mean that it is likely to freeze solid in winter and dry up completely in summer, so that many inhabitants must have arrived during the spring; how? Does the species of tree, and hence type of food resource, affect the faunal composition? Is there a pattern of succession which can be followed?

## 5. Larger bodies of water

Woodland pools and water-butts may never dry up, and hence acquire a rather different fauna. Woodland pools in particular are based on dead leaves as a food resource, are protected from wind and high evaporation rate, and the shading discourages green plants. How does the pond community developing here differ from that found in an unshaded pond?

## 6. Fallen tree trunks

Dying and dead wood provides resources for about one fifth of the woodland fauna. Some fallen trees may take as much as fifty years to decay fully. Invasion of both living or dead trees often commences with bark beetles or fungi. Some of the beetles spend their larval life on dead wood, moving to live trees after emerging from the pupa, and may in this way carry disease-causing fungus with them. It is obviously important to find out as much as possible about the beetles' own enemies to prevent the spread of damaging infections like Dutch Elm disease.

The microfungi multiplying below the loosening bark of a dead tree trunk provide the basic food supply for a succession of animal species including springtails, centipedes, millipedes, woodlice, earwigs, and beetles. What are their individual food requirements? Which feed directly on fungi and which are predators? What are the principal food chains? How do the numbers of primary consumers compare with those of their predators? Where in the tree are most of them feeding?

Many of the fungi responsible for disease and decay have characteristic and readily identifiable fruiting bodies; these include the conspicuous bracket fungi found on both living and dead trees, the dark fleshy 'Jew's Ear' (*Auricularia auricula*) on elder, and the bright cup-shaped fruiting bodies of *Peziza*.

Rotting tree stumps and fallen timber often provide a suitable place for seeds or fern spores to germinate and develop – sometimes into new trees. How many such plants are associated with your tree?

The animals associated with rotting wood may be located by dint of patient watching in the field; birds' feeding activities provide important clues. However the material is also particularly suitable for bringing back to study in the home or laboratory. A sample of rotten wood may be carefully broken up and searched in an enamel dish, or subjected to Tullgren funnel extraction (Fig. 3.13). Pieces in which wood-boring insects have been at work may be kept intact in large tins and inspected at intervals until the immature animals have completed development and emerge as adults.

## 7. Colonization of newly available surfaces

Many interesting studies of succession may be made by preparing or exposing new surfaces for colonization. Examples are cellulose-film buried in different kinds of soil (p. 124), glass or slate plates fixed on the seashore (p. 182), or in freshwater, and other new surfaces such as the hulls of boats and the piles of jetties. The factors determining the pattern of colonization may be quite subtle; juvenile stages of *Balanus balanoides* settle selectively on areas where the adult barnacles have been removed to expose bare rock, and avoid adjacent areas that were not colonized previously. Primary colonizers of all kinds provide food or shelter for a considerable range of other organisms. The undersides of stones and other objects attract a wide range of animals – woodlice on land, flatworms and freshwater-shrimps in streams, and starfish, small crabs and colonial tunicates on the lower seashore. How quickly is a new object used in this way? What kinds of behaviour bring and keep the animals there?

## 8. Other animal-derived habitats

Animal droppings and latrine areas may not only provide much useful information about the nature of the food taken (p. 221) but may serve as food resources for dung-eating animals and their predators. Some animals such as the fox and otter use their droppings as markers; once the droppings can be recognized, a study of their distribution and frequency provides useful information about the animal's activities.

Cattle dung is very important because a cow may produce up to sixteen cow-pats a day, covering well over 100 m² in a single season. A succession of organisms is associated with cow-pats; some are feeding directly on dung whilst others parasitize or prey on the dung-eaters. As the cow-pat dries at the surface or is weathered by rain, different dung-eaters and their associates become more favoured. The communities that develop in dung may be attacked by birds or mammals in search of food.

The carcases of dead animals are exploited by crows, gulls, foxes and badgers, as well as by various fly larvae, burying beetles and their associates; carrion may also be investigated by badgers and shrews seeking insects. How quickly are carrion or dung invaded? How do the invaders increase their numbers to exploit the resources in full? What effects are observable in the underlying soil?

Animal artefacts which modify the habitat and may give rise to microsuccessions include molehills and ants' nests, either of which can disturb the normal pasture flora for up to three years. Ants' nests form important food sources for insectivorous birds such as the woodpecker.

Birds' nests and their equivalents for mammals (squirrel's drey, badger sett, fox earth, rabbit hole) may yield a range of small associated animals such as ectoparasites, scavengers, and their followers. Burrows are also often associated with disturbed vegetation patterns that may be due to over-grazing, contamination with soiled bedding, or use as a latrine area. The analysis of such patterns, the determination of the roles played by nest-associates, and the identification of nesting materials, all throw light on the life of the animal concerned and may form the basis of rewarding projects.

## 9. Frontiers between ecosystems

The boundary areas between communities are particularly interesting and may draw plants or animals from either of the two habitats. Such areas include the seashore (Chapter 12), the edges of ponds and streams (p. 155), and woodland edges. Hedges correspond to woodland edge (Chapter 8) and, in Britain, form an important and accessible reservoir of woodland and other species covering an area totalling about twice that of established nature reserves. The numbers of both plant and animal species present depend partly on the soil, the age of the hedge, and the way in which the hedge has been maintained. How do hedges cut rapidly by modern mechanical means compare with similar hedges which have been cut and laid in the traditional way? How do hedges of different ages vary? Ages may be obtained from old maps, or parish records concerning tithes or enclosures. The oldest is likely to be that marking the parish boundary. Which hedge is richest in birds and shrubs? What influence does the use made of adjacent land have on hedge composition?

## 10. Community relationships

The interactions between species vary greatly, they include those of predator and prey, direct competition for particular resources, and relationships which benefit both participants. In competition for light, more rapid growth or greater ultimate height enable some plants to overshadow others; this shows most between different species, but there is enough variation within a single species to give some members an advantage. Pollination is a relationship in which two participants may benefit, the pollinating insect gaining pollen or nectar as food, and the plant being enabled to reproduce. In such mechanisms, other participants may be involved in unexpected ways, as in Darwin's explanation of why red clover grew more vigorously near villages; red clover is pollinated by bumble bees, bumble bees nests are eaten by field mice – but field mice are eaten by cats who live in villages, so that bumble bees and red clover are better able to flourish there. What factors affect the numbers of other pollinating insects?

Another kind of mutual benefit may be found in the relationship between ants and aphids. Ants get honeydew from aphids, but may help protect the aphid population in a variety of ways. How do aphid populations from which ants have been experimentally excluded compare with those associated with ants ?

Animal colouration affects relationships because of its value in camouflage and in warning of distastefulness and danger. Harmless insects may avoid attack because of a resemblance to more dangerous animals, as in the many different species which mimic bees and wasps. What other examples are there ? How do the numbers of mimics compare with those of the 'model' they are copying ?

Other kinds of response are brought into play when a new plant is transplanted into a community, and this may be particularly instructive where it will be competing with a closely related native species. What advantages, if any, does the native plant have over the plant from a different habitat ? What factors make for success or failure in the new environment ?

There are many reasons why one species may be more competitive than another. These include reproductive efficiency, adaptations for dispersal, and devices which discourage attack by other species – such as spines on gorse or hedgehogs, or unpleasant taste of buttercups or brightly coloured caterpillars. How do similar pairs of dispersal mechanisms such as dandelion fruits and thistledown, or sycamore and ash fruits, compare in efficiency ?

## 11. Habitat factors

The effects of habitat factors may be investigated by comparing existing sites, or by experimental interference. Shade tolerance and the effect of light on plant stature or flower colour might be investigated by direct observation; the conclusions could then be tested by artificially altering the amount of shade reaching similar plants. Near main roads in urban areas, the continuous artificial light during hours of darkness could be important to animals by providing an extended twilight. What effect does street lighting have on animal behaviour ? Are plants also affected ?

Many other investigations, involving such factors as temperature, humidity, soil factors, water flow rates, trampling, population density, may be pursued similarly by both direct observation and experiment.

## 12. Distribution studies

The methods outlined in Chapters 3 and 4 may be used for recording the distribution of animals and plants. Such records should include not only the distribution at one moment in time, but how the pattern varies over an extended period. What relationships are there between the distributions of consumers and consumed? In woodland, vertical distribution should be investigated using a combination of sticky-traps and collecting nets for insects, and direct observation for the birds that feed on them.

Profitable comparisons may be made between habitats that are managed in different ways. How do the numbers of both animal and plant species compare in grazed (or mown) and ungrazed grassland? Insect numbers can be compared by collecting in a standardized way with a net or pooter for a fixed time in each habitat.

Transects should sample animals as well as plants, possibly by the regular placing of pitfall traps (Fig. 3.6) and small mammal traps.

### 13. Energy flow* (Added at 1979 reprint)

Energy flow through a community may be studied by first measuring the input of plant material per unit area into the community (pp. 66, 133) and then sampling the herbivore, carnivore and decomposer populations supported by a given area, and assessing their use of the available material. Woodland leaf litter is a particularly suitable subject for this kind of study. The energy budgets of individual animals can be determined in the laboratory by measuring three out of the four parameters: consumption $(C)$, production $(P)$, respiration $(R)$, waste – faeces and excretion $(W)$. The relationship between them is: $C - W = P + R$. For a herbivore, consumption can be measured directly by weighing plant material before and after the animal has been feeding on it for a set time. As *dry* weight is required, and there is likely to be some loss of water by evaporation during the time of feeding, a separate determination of fresh to dry weight ratio should be made at the time of setting up the experiment; from this, the original dry weight of the plant material can be calculated. Consumption is then the difference between this value and the dry weight of the uneaten plant material. Droppings should be collected, dried and weighed. Some ingenuity will be needed to devise a suitable method for consumption and egestion for any particular carnivore or detritus eater.

Gaseous exchange during respiration can be determined using capillary-tube gas analysis (p. 121f.). This will provide information about rate of oxygen consumption and Respiratory Quotient, where

$$R.Q. = \frac{\text{volume of } CO_2 \text{ given off}}{\text{volume of } O_2 \text{ taken up}}.$$

Respiration may be expressed in joules on the basis that $1 \text{ cm}^3 \text{ O}_2$

---

* Grateful acknowledgement to Dr R. J. Putman, University of Southampton, for ideas and data used here.

TABLE 15.1—TABLE OF APPROXIMATE CALORIFIC VALUES

| Material | kJ/g dry wt | kJ/g fresh wt |
|---|---|---|
| Grass | 18.6 | 5.6 |
| Leaf litter | 17.8 | |
| Fruits and seeds | 21.2 | |
| *Cabbage | 18.3 | |
| Insects (general) | 22.4 | |
| *Caterpillars (*Pieris brassicae*) | 19.7 | |
| *Caterpillar faeces | 13.25 | |
| Annelids | 19.3 | |
| Molluscs without shell | 19.2 | |
| Molluscs with shell | 15–16 | |
| Laboratory mice | 24.6 | 9.7 |
| Wild rodents | | 6.3 |
| *Mouse faeces | 18.0 | |
| *Mouse pellets (laboratory diet) | 20.9 | |
| White bread | 16.3 | 10.6 |

* Figures concerning animals reared in the laboratory – enabling the technique to be tried and tested before use with wild animals.

consumed represents $21.1 \times$ R.Q. joules of energy expended ($= 5.05 \times$ R.Q. calories).

When measurements are complete, reference to Table 15.1 will give *approximate* energy values for each trophic level. Calorific values of some foods or droppings may be estimated in the laboratory with a food calorimeter as used in Nuffield courses, by comparing the values obtained with a known standard (from food tables: e.g. 1 g of dry white bread yields approximately 16.3 kilojoules on combustion).

Complete energy budgets ($C - W = P + R$) can be drawn up for some of the animals. Assimilation efficiencies ($(C - W)/C \times 100$)% should be compared between trophic levels, and consideration should also be given to efficiency of energy transfer from one level to another

$$\left( \frac{\text{Energy taken in by trophic level } n}{\text{Energy taken in by trophic level } (n-1)} \times 100 \right)\%.$$

Finally, the total flow of energy through the community may be investigated in two ways:

(i) a comparison of the energy input in plant material with the energy consumed at different trophic levels. The energy consumption of herbivores is the sum total of *biomass × energy consumed per gram animal* (from $C$ in energy budget, above) for all herbivore species. Similar determinations should be made for carnivores and detritivores.

(ii) in a stable community with constant biomass, no energy is being stored by biomass accumulation or released by overall loss.

Energy entering the system will then match the energy lost in respiration by plants, herbivores, carnivores and detritivores:

*Total energy input in plant material = total energy output in respiration*

This list of topics for investigation is by no means complete, but we hope the many questions asked may prompt you to ask and find the answers to similar questions in your own particular study area.

# Appendix

## A RECIPES FOR FUNGAL CULTURE

1. *Oatmeal agar*

    | | |
    |---|---|
    | Oatmeal | 30 g |
    | Agar | 20 g |
    | Water | 1 litre |

    30 g of powdered oatmeal is boiled in a double saucepan in water; squeeze through muslin and make up to 1 litre. Boil with the agar, stirring until the agar is dissolved. Autoclave (heat in a pressure cooker) in tubes or petri dishes at 15 lb. for 20 min.

2. *Potato dextrose agar*

    | | |
    |---|---|
    | Potato | 200 g |
    | Dextrose | 20 g |
    | Agar about | 20 g |
    | Water | 1 litre |

    Cut the potato in small pieces, cover with water and boil gently for half an hour (or pressure cook for 10 min). Cool and allow to settle; pour off supernatant liquid and make up to 1 litre. Add the dextrose and agar, heating and stirring until dissolved. Autoclave as above.

3. *Soil, dung or hay infusion agar*

    | | |
    |---|---|
    | Fertile soil | 200 g |
    | *or* Horse dung | 200 g |
    | *or* Chopped hay | 25 g |
    | Water | 500 cm³ |
    | Agar | 7.5 g |

    Boil the soil, dung or hay gently in the water for one hour. Then filter, and make up to 500 cm³. Add the agar with stirring and controlled heat. Autoclave as before.

4. *Other Media*

    Many media are available in tablet and powder form from Oxoid Ltd. These include Malt Agar and Czapek Dox Agar (for fungi) and Nutrient Agar and Blood Agar (for bacteria).

## B NIGROSINE AND PICRO-LACTOPHENOL

| | |
|---|---|
| Phenol crystals | 10 g |
| Picric acid (saturated solution) | 10 cm³ |
| Lactic acid | 10 cm³ |
| Glycerol | 20 cm³ |

Add phenol crystals to picric acid and warm until dissolved. Add lactic acid and glycerol.

Nigrosine 2% aqueous solution.

Add nigrosine to picric lactophenol just before use, using equal parts of each.

## C ADDITIONAL NOTES ON WATER ANALYSIS

The most useful sources for methods are:

MACKERETH, F. J. H., *Some Methods of Water Analysis for Limnologists.* Scientific Publication No. 21, Freshwater Biological Association.
DOWDESWELL, W. H., *Practical Animal Ecology.* Methuen.

A comprehensive guide to methods for estimating dissolved substances is given in *Chemical methods of water testing*, published by The British Drug Houses Ltd., B.D.H. Laboratory Chemicals Division, Poole, Dorset, who also offer advice through their Technical Services Department at the same address.

The following notes on oxygen determination, organic matter content, and B.O.D., should prove useful when investigating organic pollution.

*Collecting samples*

Reagent bottles of 250 cm³ capacity are best for collecting samples. Always take the water temperature. Because air already in the bottle can affect gases dissolved in the sample, it is essential to avoid agitation and

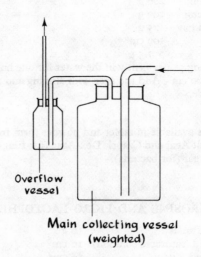

Overflow vessel

Main collecting vessel (weighted)

**Fig. C.1** Sampling bottle

bubbling, and to arrange for the initial inflow to be displaced. Figure C.1 shows a sampling bottle which fills slowly from the bottom through a wide inlet tube whilst air, then water, are expelled through a narrower tube into a second vessel which has its outlet well above the top of the original inlet – so ensuring that the apparatus fills in the right direction.

Suitably weighted, or fixed to a calibrated rod, this apparatus can take samples at chosen depths by jerking a string to pull a bung from the final outlet. Immediately after collection, the rubber bung of the sample bottle should be replaced by a ground-glass stopper, taking care not to trap an air bubble.

Samples from very shallow waters may be collected using a 100 cm$^3$ pipette, then quickly and gently transferring to a 60 cm$^3$ bottle, allowing the excess to overflow.

### Determination of dissolved oxygen

#### (a) Winkler method

Immediately after the sample has been taken, add 2 cm$^3$ of 50% manganous chloride solution and 2 cm$^3$ of Winkler's reagent (100 g potassium hydroxide and 60 g potassium iodide in 200 cm$^3$ water) using pipettes* that reach well below the surface. This may displace some water from the top. Replace the stopper, excluding air, and mix thoroughly by vigorously inverting and rotating the bottle. A precipitate of manganous hydroxide forms, but oxygen in the water will convert some of this into a corresponding amount of manganic hydroxide.

On return to the laboratory, allow the precipitate to settle. Then introduce* carefully 2 cm$^3$ of concentrated sulphuric acid, replacing the stopper quickly and avoiding the introduction of air or loss of precipitate. Mix thoroughly by rotation, dissolving the precipitate; the manganic hydroxide oxidizes potassium iodide to an equivalent amount of free iodine. This means that the concentration of dissolved oxygen may be estimated by determining the concentration of iodine set free. A rough estimate can be made from the depth of colour of the iodine. More accurate determination may be made by comparison with glass standards in a B.D.H. comparator, or by titration.

Titration: Rotate to mix; then, using a pipette, transfer 100 cm$^3$ from the sample bottle into a conical flask. Without delay titrate with standardized N/80 sodium thiosulphate solution until only a faint yellow colour remains; add a few drops of freshly prepared starch solution, then add more thiosulphate drop by drop until the blue colour just disappears. Each 1 cm$^3$ of thiosulphate used is equivalent to 1 mg of oxygen per litre; thus $x$ cm$^3$ of thiosulphate indicate an oxygen concentration of $x$ mg per litre.

NOTE: This method is said to be unreliable for sea water.

*CAUTION: These reagents must not be sucked into pipettes by mouth; use bulb pipettes or dipping pipettes.

*(b) Phenosafranine method for use in the field*

(b) (Ref.: DOWDESWELL, W. H., *Practical Animal Ecology*. Methuen Nuffield Advanced Science. BIOLOGY. *Maintenance of the Organism*, Penguin.)

To protect the sample from atmospheric oxygen during titration, introduce a few drops of olive oil or liquid paraffin into a boiling tube; then add gently 50 cm$^3$ of the sample underneath the oil with a pipette. If it is impractical to use a pipette in the field, fill a plastic measuring cylinder or calibrated boiling tube with sample water to the 50 cm$^3$ mark, avoiding disturbance which might introduce oxygen; then add the few drops of oil. Introduce in turn the following liquids:

(1) 5 cm$^3$ of Fehling's solution B (173 g sodium potassium tartrate and 120 g sodium hydroxide per litre)

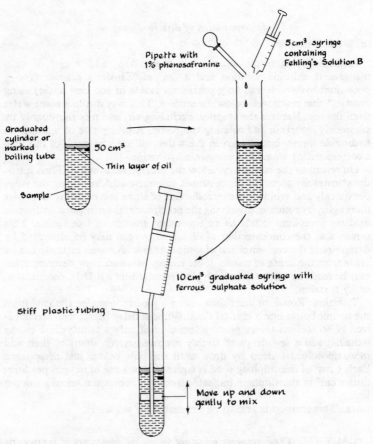

Pipette with
1% phenosafranine

5 cm$^3$ syringe
containing
Fehling's Solution B

Graduated
cylinder or
marked
boiling tube

50 cm$^3$

Thin layer of oil

Sample

10 cm$^3$ graduated syringe with
ferrous sulphate solution

Stiff plastic tubing

Move up and down
gently to mix

**Fig. C.2** Oxygen determination using phenosafranine

(2) About 2 drops of 1% aqueous phenosafranine solution (sufficient to colour the solution pink)

(3) *Freshly prepared* ferrous sulphate solution containing 2.48 g of $FeSO_4.7H_2O$ (Analar standard) in 1 litre of 1% $H_2SO_4$. Use a specially prepared (Fig. C.2) 10 $cm^3$ syringe (or pipette) with a rubber piston on the outlet, to add ferrous sulphate solution a little at a time and gently mix it with the sample until the pink colour disappears. The piston may be a suitable tap-washer or cut from a rubber bung, and must be kept below the surface during the entire operation so that no extra air is drawn in.

The volume in $cm^3$ of ferrous sulphate required to make the solution colourless is equivalent to the weight in mg of dissolved oxygen per litre. In other words, $x$ $cm^3$ $FeSO_4$ solution indicates that the sample contains

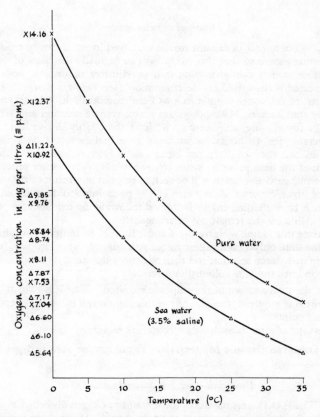

**Fig. C.3** Oxygen in saturated solutions at normal pressure (Figures from: Truesdale, G. A., Downing, A. L., and Lowden, G. F. (1955). The solubility of oxygen in pure water and sea water. *J. appl. Chem.*, 5, 53–62)

$x$ mg $O_2$ per litre. The titration should be repeated at least twice for each sample to ensure that the result is reliable.

This method may be used to compare the oxygen contents in a set of samples. However, for accurate oxygen determination, the result must be corrected by multiplying by a factor which has been derived by comparing the experimental value for a saturated oxygen solution with the correct value. A saturated solution is obtained by bubbling air through, or by shaking water vigorously with air. Determine the oxygen value using the method described above. From the graph (Fig. C.3) determine the correct value for a saturated oxygen solution at the temperature of the experiment. Then:

$$\text{The correcting factor } F = \frac{\text{correct value for saturated } O_2 \text{ soln.}}{\text{experimental value for saturated } O_2 \text{ soln.}}$$

Correct $O_2$ value for sample $= F \times x$ mg $O_2$ per litre.

### Dissolved organic matter

Much, but not all, of organic matter dissolved in water can be oxidized by permanganate so that this oxidation can be made the basis of determinations which can give comparative estimates of organic pollution more rapidly than B.O.D. determination (see below). *Method:* Place 100 cm³ of the water sample in a 250 cm³ conical flask. Add 10 cm³ of freshly standardized N/80 potassium permanganate solution and 10 cm³ of 25% (by volume) sulphuric acid. Shake thoroughly. *Either* leave in a refrigerator for 12 hours, *or* incubate in a boiling water bath for 30 minutes and then cool. In the latter case a filter funnel placed in the mouth of the flask prevents losses from splashing. After either procedure the permanganate solution will be seen to be paler as a result of the oxidation of organic matter. Add 2 cm³ of 5% potassium iodide solution, and shake. A brown iodine colour is formed following the oxidation of potassium iodide by the remaining permanganate.

Titrate this iodine with freshly standardized N/80 sodium thiosulphate solution until only a faint yellow colour remains; add a few drops of freshly prepared starch solution and then add more thiosulphate solution drop by drop until the blue colour disappears.

NOTE: the total amount of thiosulphate required will be less than 10 cm³, so caution is required *from the start* of this titration; it is easy to overshoot the end point.

If a total of $x$ ml thiosulphate solution are required:

The Dissolved Organic Matter $\equiv (10 - x)$ mg oxygen absorbed per litre.

### Biochemical oxygen demand (B.O.D.)

(1) The B.O.D. test measures the amount of oxygen absorbed biologically by the sample in the course of 5 days at 20°C. Much of the oxygen is used to oxidize organic carbon and some may be used in nitrification of organic nitrogen compounds or ammonia. The test therefore reflects the

ecological quality of polluted water better than the permanganate test which may oxidize materials not normally oxidized biologically in rivers. However, a B.O.D. much lower than that predicted from the permanganate test may indicate the presence of toxic effluent. With even moderate organic pollution, the sample may be capable of absorbing more oxygen in 5 days than could normally be present at the start of the experiment, so that it may be necessary to dilute the sample with well oxygenated water first (see note). The dilution should be so arranged that not more than 70% of the oxygen present immediately after dilution has disappeared by the end of 5 days. The amount of dilution needed will vary with the degree of pollution up to about 1 in 10, but greater dilution may be needed for raw sewage and some industrial effluents. At the start it is probably best to examine several dilutions (e.g. pure sample, 1 in 2, 1 in 5, 1 in 10) and reject those results which show oxygen absorption exceeding 70%. With experience, an appropriate dilution can be selected in the light of a previous determination of dissolved organic matter with permanganate.

Oxygen determinations are made using either of the two methods outlined above (Winkler, and phenosafranine). One set of full strength and diluted samples is tested immediately after collection. Another set is kept in full stoppered 250 cm³ bottles and incubated in a water bath at 20°C for 5 days before determining the oxygen content.

If the original sample contained $x$ mg of oxygen per litre and the incubated sample $y$ mg of oxygen per litre, then:

'5 day B.O.D.' = $(x - y)$ mg oxygen per litre.

If diluted samples are used the calculated figure must be multiplied by the dilution factor to give the right B.O.D.

NOTE: Dilution water must be free of oxidizable matter and substances which might inhibit bacterial growth, but distilled water is unsuitable. In many cases water taken from a stream above the polluting source may be used, but its own B.O.D. should be determined and applied as a correction. Artificial dilution water is not easily prepared, and is described in *Analysis of Raw, Potable and Waste Waters* (H.M.S.O., 1972).

(2) A good and simple way of *comparing* B.O.D. values in a set of samples is as follows:

(a) Collect samples as before in 250 cm³ bottles with ground-glass stoppers.

(b) Using a pipette that reaches well below the surface, add 1 cm³ of 0.1% methylene blue solution to each sample.

(c) Replace the stoppers immediately without introducing air.

(d) Transfer the bottles to a warm (about 20°C) dark place, noting the time; examine at intervals for disappearance of the blue colour.

Methylene blue remains blue as long as oxygen is present, but goes colourless when the oxygen has been used up; the times taken for the colour to disappear therefore provide a basis for comparing the rates at which oxygen is being used up in the samples.

# Further Reading

Useful sources of additional information are given under appropriate headings below. Local information may be available in the form of regional floras and faunas such as the *Flora of Gloucestershire* and the *Plymouth Marine Fauna*, or in books of the Regional Naturalist Series published by David and Charles. These often provide a useful check on provisional identification because they describe and name the places where species have been found; if your species is in an atypical habitat, it is obviously necessary to reconsider your identification. Many publications of scientific bodies contain useful accounts of ecological surveys and methods. These publications include:

*Association for Science Education*
College Lane, Hatfield, Herts.

School Science Review
Science Masters' Book, Series IV, Part III
Lab Books (No. 3, Ecology)

*British Ecological Society*
Harvest House, 62 London Road, Reading, RG1 5AS.

Journal of Ecology (mostly botanical, including the Biological Flora of the British Isles, available as reprints)
Journal of Animal Ecology
Journal of Applied Ecology
Proceedings of various symposia

*British Trust for Ornithology*
Beech Grove, Tring, Herts.

Bird Study
Guides (various)

*Mammal Society of the British Isles*
c/o The Institute of Biology.

Mammal Review

*Institute of Biology*
41 Queen's Gate, London SW7 5HU.

Journal of Biological Education
Studies in Biology (several titles), Edward Arnold

*General books on ecological topics:*

ASHBY, M. (1969). *Introduction to Plant Ecology*, 2nd edn. London: Macmillan.

BISHOP, O.N. (1973). *Natural Communities*. John Murray.

BISHOP, O. N. (1975). *Beginning Field Biology*. Harrap.

CLARK, E. (1973). *Fieldwork in Biology – an Environmental Approach*. Macmillan.

DAJOZ, R. (1977). *Introduction to Ecology*. Hodder & Stoughton.

DOWDESWELL, W. H. (1966). *Introduction to the Ecology of Animals*. Methuen.

ELTON, C. S. (1958). *The Ecology of Invasions by Animals and Plants*. Methuen.

ELTON, C. S. (1966). *Animal Ecology*. Science Paperback.

ELTON, C. S. (1966).  *The Pattern of Animal Communities.* Methuen.
ELTON, C. S. (1967).  *The Ecology of Animals.* Chapman and Hall.
FORD, E. B. (1974).  *Ecological Genetics.* Methuen.
GRAHAM, M. (1973).  *A Natural Ecology.* Manchester University Press.
KORMONDY, E. J. (1969).  *Concepts of Ecology.* Prentice-Hall.
KREBS, C. J. (1972).  *Ecology; the Experimental Analysis of Distribution and Abundance.* Harper & Row.
LACK, D. (1970).  *The Natural Regulation of Animal Numbers.* Clarendon Press.
ODUM, E. P. (1963).  *Ecology.* Holt, Rinehart & Winston (Modern Biology Series).
ODUM, E. P. (1971).  *Fundamentals of Ecology,* 3rd Edn. Saunders.
PHILLIPSON, J. (1966).  *Ecological Energetics* (Studies in Biology No. 1). Edward Arnold.
RICKLEFS, R. E. (1973).  *Ecology.* Nelson.
TANSLEY, SIR A. G. (1950).  *The British Islands and their Vegetation.* Cambridge University Press.

Most titles in Collins' *New Naturalist Series* will be of interest and value to ecologists.

*Collecting and Recording:*

JENNINGS, T. J. (1971).  *Collecting from Nature.* Wheaton.
NATURAL HISTORY MUSEUM (1963; 1954; 1956).  *Instructions for Collectors*: No. 4A, *Insects*; No. 9A, *Invertebrates other than Insects*; No. 10, *Plants.* H.M.S.O.
OLDROYD, H. (1970).  *Collecting, Preserving and Studying Insects.* Hutchinson.
WAGSTAFFE, R., and FIDLER, J. H. (1955).  *The Preservation of Natural History Specimens.* Vols. I & II. Witherby.

Publications of the Amateur Entomologists' Society, 23 Manor Way, North Harrow, Middlesex.

*Methods and Experimental Design:*

ANDREWARTHA, H. G. (1961).  *Introduction to the Study of Animal Populations.* Methuen.
DAVIES, A., Ed. (1973).  *Ecology*; A.S.E. Lab Book 3. John Murray.
DOWDESWELL, W. H. (1959).  *Practical Animal Ecology.* Methuen.
HEATH, O. V. S. (1970).  *Investigation by Experiment* (Studies in Biology No. 23). Edward Arnold.
LEWIS, T. and TAYLOR, L. R. (1967).  *Introduction to Experimental Ecology.* Academic Press Inc.
MACFADYEN, A. (1970).  *Animal Ecology: Aims and Methods.* Pitman.
MACKERETH, F. J. H., HERON, J. and TALLING, J. F. (1978).  *Water Analysis – Some Revised Methods for Limnologists* (Scientific Publication No. 36). Freshwater Biological Association.
SOUTHWOOD, T. R. E. (1966).  *Ecological Methods, with Particular Reference to the Study of Insect Populations.* Methuen.

WADSWORTH, R. M., Ed., (1968). *The Measurement of Environmental Factors in Terrestrial Ecology* (B.E.S. Symposium Proceedings). Blackwell.

In addition, the International Biological Programme Handbooks, published by Blackwell Scientific Publications and planned to give guidance to biologists participating in the world programme, give detailed recommendations about methods. Some suggestions may be beyond the scope of a student with limited resources, but the discussions are always illuminating and helpful in assigning priorities when planning an investigation.

*Identification:*

There are a number of series of publications which are generally useful. Most important of these for beginners are:

| | |
|---|---|
| *Blandford Colour* series | Blandford |
| *Collins' Guide* series | Collins |
| *Observer's Book* series | Warne |
| *Oxford Book* series | O.U.P. |
| *Penguin Nature Guides* | Penguin |
| *Wayside and Woodland* series | Warne |
| *Young Specialist* series | Burke |

Particularly useful books from these series are mentioned below or under habitat headings.

CLAPHAM, A. R., TUTIN, T. G., and WARBURG, E. F. (1968). *Excursion Flora of the British Isles.* Cambridge University Press.

CLOUDSLEY-THOMPSON, J. L. and SANKEY, J. (1961). *Land Invertebrates.* Methuen.

CORBET, G. B. (1964). *The Identification of British Mammals.* British Museum (Natural History).

CORBET, G. B. and SOUTHERN, H. N. Eds. (1977). *The Handbook of British Mammals,* 2nd edition. Blackwell.

DAGLISH, E. G. (1952). *Name this Insect.* Dent.

EDLIN, H. L. (1975). *Know Your Broadleaves* (Forestry Commission Publication 20). H.M.S.O.

EDLIN, H. L. (1976). *Know Your Conifers* (Forestry Commission Publication 15). H.M.S.O.

FITTER, R. S. R. and MCCLINTOCK, D. (1956). *Collins Pocket Guide to Wild Flowers.* Collins.

HUBBARD, C. E. (1968). *Grasses.* Penguin (Pelican series).

HYDE, H. A., WADE, A. E., and HARRISON, S. G. (1969). *Welsh Ferns, Clubmosses, Quillworts, and Horsetails.* National Museum of Wales (Cardiff).

   (Describes all pteridophytes native to the British Isles and the Channel Islands as well as some naturalized aliens.)

JERMY, A. C. and TUTIN, T. G. (1968). *British Sedges.* Botanical Society of the British Isles.

KEBLE MARTIN, W. (1969). *The Concise British Flora in Colour.* Ebury Press/Michael Joseph, and Sphere Books (paperback).

LANGE, M. and HORA, F. B. (1965). *Collins Guide to Mushrooms and Toadstools.* Collins.

MEIKLE, R. D. (1958).   *British Trees and Shrubs*. Eyre & Spottiswoode (*Kew* series).

PAVIOUR-SMITH, K., and WHITTAKER, J. B. (1968).   *A Key to the Major Groups of British Free-Living Terrestrial Invertebrates*. Blackwell.

PETERSON, R., MOUNTFORD, G., and HOLLOM, P. A. D. (1973).   *Field Guide to the Birds of Britain and Europe*. Collins.

PHILIPS, R. (1977).   *Wild Flowers of Britain*. Pan Books.

SMITH, M. (1951).   *British Amphibians and Reptiles*. Collins (New Naturalist Series).

WATSON, E. V. (1968).   *British Liverworts and Mosses*. Cambridge University Press.

For more advanced identification there are specialist works published by:

The Freshwater Biological Association, The Ferry House, Ambleside, Westmorland (keys for identifying freshwater animals, and methods of water analysis).

The Linnaean Society of London, Burlington House, Piccadilly, London, WiV OLQ (synopses of the British fauna).

The Royal Entomological Society, 41 Queen's Gate, London, SW7 5HU (handbooks for the identification of British insects).

Seashore and freshwater works are listed under the appropriate headings below.

*Quantitative Studies:*

*(a) General*

ANDREWARTHA, H. G. (1970).   *Introduction to the Study of Animal Populations*. Methuen.

GRIEG-SMITH, P. (1957).   *Quantitative Plant Ecology*. Butterworth.

KERSHAW, K. A. (1973).   *Quantitative and Dynamic Plant Ecology*, 2nd edn. Edward Arnold.

LEWIS, T., and TAYLOR, L. R. (1967).   *Introduction to Experimental Ecology*. Academic Press.

SOLOMON, M. E. (1976).   *Population Dynamics*, 2nd edition (Studies in Biology 18). Edward Arnold.

*(b) Statistics*

BAILEY, N. T. J. (1959).   *Statistical Methods in Biology*. English Universities Press.

BISHOP, O. N. (1966).   *Statistics for Biology*. Longmans.

CLARKE, G. M. (1969).   *Statistics and Experimental Design*. Edward Arnold.

MORONEY, M. J. (1951).   *Facts from Figures*. Penguin (Pelican series).

PARKER, R. E. (1973).   *Introductory Statistics for Biology* (Studies in Biology No 43). Edward Arnold.

*Soil:*

CLOUDSLEY-THOMPSON, J. L. (1967).   *Microecology* (Studies in Biology No. 6). Edward Arnold.

DARLINGTON, A. and SMYTH, J. C. (1966). *Keys to Small Organisms in Soil, Litter and Water Troughs.* Longmans/Penguins for Nuffield Foundation.

FITZPATRICK, E. A. (1974). *An Introduction to Soil Science.* Oliver & Boyd.

JACKSON, R. M. and RAW, F. (1966). *Life in the Soil* (Studies in Biology No 2). Edward Arnold.

KEVAN, D. K. MCE. (1968). *Soil Animals.* Witherby.

LEADLEY-BROWN, A. (1978). *Ecology of Soil Organisms.* Heinemann Educational.

PHILLIPSON, J. (1971). *Methods of Study in Quantitative Soil Ecology; Population, Production, and Energy Flow* (I.B.P. Handbook No. 18). Blackwell.

RUSSELL, SIR E. J. (1959). *The World of the Soil.* Collins (New Naturalist series). (Also published 1970 by Fontana Books).

RUSSELL, E. W. (1961). *Soil Conditions and Plant Growth.* Longmans.

SHEPLEY, A. (1973). *Soil Studies.* Pergamon.

TRIBE, H. T. (1967). *Practical studies on biological decomposition in soil: a simple technique for observation of soil organisms colonizing buried cellulose film,* School Science Review No. 167. John Murray.

WALLWORK, J. A. (1970). *Ecology of Soil Animals.* McGraw-Hill.

## Woodland:

COUSENS, J. (1974). *An Introduction to Woodland Ecology.* Oliver & Boyd.

DARLINGTON, A. (1968). *Plant Galls in Colour.* Blandford.

MANDAHL-BARTH, G. (1966). *Woodland Life in Colour.* (Translated by A. Darlington). Blandford.

NEAL, E. G. (1958). *Woodland Ecology.* Heinemann.

OVINGTON, J. D. (1965). *Woodlands.* English Universities Press.

PRIME, C. (1970). *Investigations in Woodland Ecology.* Heinemann (Investigations in Biology series).

## Grassland:

LOUSLEY, J. E. (1951). *Wild Flowers of Chalk and Limestone.* Collins (NN).

LYNEBORG, L. (1968). *Field and Meadow Life in Colour.* (Translated by A. Darlington). Blandford.

MOORE, I. (1966). *Grass and Grasslands.* Collins (New Naturalist series).

SALISBURY, SIR E. J. (1952). *Downs and Dunes.* Bell.

SANKEY, J. (1966). *Chalkland Ecology.* Heinemann.

SPEDDING, C. R. W. (1971). *Grassland Ecology.* Clarendon Press.

## Heath, Moorland and Bog:

FRIEDLANDER, C. P. (1960). *Heathland Ecology.* Heinemann.

GIMINGHAM, C. H. (1972). *Ecology of Heathlands.* Chapman & Hall.

GIMINGHAM, C. H. (1975). *An Introduction to Heathland Ecology.* Oliver & Boyd.

PEARSALL, W. H. (1950, revised 1971). *Mountain and Moorland.* Collins (New Naturalist series).

*Freshwater:*

CLEGG, J. (1967).   *Observer's Book of Pond Life.* Warne.

DOWDESWELL, W. H., Ed., (1970).   *Key to Pond Organisms* (Nuffield Advanced Science). Penguin.

ENGELHARDT, W. (1964).   *Pond Life.* Burke (Young Specialist series).

GARNETT, W. J. (1965).   *Freshwater Microscopy.* Constable.

HYNES, H. B. (1970).   *The Ecology of Running Waters.* Liverpool University Press.

LEADLEY BROWN, A. (1971).   *The Ecology of Fresh Water.* Heinemann.

MACAN, T. T. (1959).   *A Guide to Freshwater Invertebrate Animals.* Longmans.

MACAN, T. T. (1963).   *Freshwater Ecology.* Longmans.

MACAN, T. T. (1973).   *Ponds and Lakes.* Allen and Unwin.

MACKERETH, F. J. H., HERON, J. and TALLING, J. F. (1978).   *Water Analysis – Some Revised Methods for Limnologists* (Scientific Publication No. 36). Freshwater Biological Association.

MELLANBY, H. (1963).   *Animal Life in Fresh Water.* Methuen.

MILLS, D. H. (1972).   *An Introduction to Freshwater Ecology.* Oliver and Boyd.

NEEDHAM, J. G. and NEEDHAM, P. R. (1963).   *A Guide to the Study of Freshwater Biology.* Constable.

POPHAM, E. J. (1955).   *Some Aspects of Life in Fresh Water.* Heinemann.

QUIGLEY, M. (1977).   *Invertebrates of Streams and Rivers.* Edward Arnold.


*The Seashore and Maritime Habitats:*

BARNES, R. S. K. (1974).   *Estuarine Biology* (Studies in Biology 49). Edward Arnold.

BARRETT, J. H. and YONGE, C. M. (1958).   *Pocket Guide to the Seashore.* Collins.

BONEY, A. D. (1975).   *Phytoplankton* (Studies in Biology 52). Edward Arnold.

BRAFIELD, A. E. (1978).   *Life in Sandy Shores* (Studies in Biology 89). Edward Arnold.

CHAPMAN, V. J. (1964).   *Coastal Vegetation.* Pergamon.

EALES, N. B. (1967).   *The Littoral Fauna of Great Britain,* 4th edn. Cambridge.

ELTRINGHAM, S. K. (1971).   *Life in Mud and Sand.* English Universities Press.

EVANS, S. M. and HARDY, J. M. (1970).   *Seashore and Sand Dunes.* Heinemann (Investigations in Biology series).

GREEN, J. (1968).   *The Biology of Estuarine Animals.* Sidgwick & Jackson.

HAAS, W. DE and KNORR, F. (1966).   *Marine Life.* Burke (Young Specialist series).

HARDY, SIR A. C. (1956).   *The Open Sea: I The World of Plankton.* Collins (New Naturalist series).

HARDY, SIR A. C. (1959).   *The Open Sea: II Fish and Fisheries.* Collins (New Naturalist series).

KOSCH, A., FRIELING, H. and JANUS, H. (1963).   *Seashore.* Burke (Young Specialist).

LEWIS, J. R. (1964). *The Biology of Rocky Shores.* English Universities Press.

MCLUSKY, D. S. (1971). *Ecology of Estuaries.* Heinemann.

NEWELL, G. E. (1964). *Plankton.* Hutchinson.

NEWTON, L. (1931). *A Handbook of the British Seaweeds.* H.M.S.O.

RANWELL, D. S. (1972). *Ecology of Salt Marshes and Sand Dunes.* Chapman and Hall.

SALISBURY, SIR E. J. (1952). *Downs and Dunes.* Bell.

SOUTHWARD, A. J. (1965). *Life on the Seashore.* Heinemann.

WICKSTEAD, J. H. (1976). *Marine Zooplankton* (Studies in Biology 62). Edward Arnold.

WILSON, D. P. (1951). *Life of the Shore and Shallow Sea.* Nicholson and Watson.

YONGE, C. M. (1963). *The Sea Shore.* Collins (New Naturalist series).

## Projects

Many of the references given above should provide ideas for project work and problems to investigate. The following should also prove useful:

*General and Community Studies:*

CHINERY, M. (1977). *The Natural History of the Garden.* Collins.

COTT, H. B. (1940). *Adaptive Coloration in Animals.* Methuen.

DALE, A. (1960). *Observations and Experiments in Natural History.* Heinemann.

DARLINGTON, A. (1969). *Ecology of Refuse Tips.* Heinemann.

DENNIS, E., Ed. (1972). *Everyman's Nature Reserve: Ideas for Action.* David and Charles.

DEVON TRUST FOR NATURE CONSERVATION. (1972). *School Projects in Natural History.* Heinemann.

ENNION, E. A. R. (1958). *Bird Studies in a Garden.* Penguin (Puffin series).

SALISBURY, SIR E. J. (1942). *The Reproductive Capacity of Plants.* Bell.

SALISBURY, SIR E. J. (1961). *Weeds and Aliens.* Collins New Naturalist series).

SAVORY, T. H. (1955). *The World of Small Animals.* University of London Press.

TAUNTON, J. (1969). *Bird Projects for Schools.* (Royal Society for the Protection of Birds.) Evans Bros.

WELTY, J. C. (1964). *The Life of Birds.* Constable.

WILSON, R. W. and WRIGHT, D. F. (1972). *A Field Approach to Biology: Teachers' Guide.* Heinemann.

YAPP, W. B. (1970). *The Life and Organization of Birds.* Edward Arnold.

*Behaviour Studies:*

CARTHY, J. D. (1956). *Animal Navigation.* Allen & Unwin.

CARTHY, J. D. (1958). *An Introduction to the Behaviour of Invertebrates.* Allen & Unwin.

CARTHY, J. D. (1966). *The Study of Behaviour* (Studies in Biology No. 3). Edward Arnold.

CROWFOOT, P. (1966).  *Mice All Over*. G. T. Foulis.

EVANS, S. M. (1968).  *Studies in Invertebrate Behaviour*. Heinemann.

EVANS, S. M. (1970).  *The Behaviour of Birds, Mammals, and Fish*. Heinemann.

HOWARD, L. (1952).  *Birds as Individuals*. Collins.

HOWARD, L. (1956).  *Living with Birds*. Collins.

KLOPFER, P. H. (1973).  *Behavioural Aspects of Ecology*. Prentice-Hall.

LACK, D. (1951).  *The Life of the Robin*. Penguin (Pelican series).

LORENZ, K. (1961).  *King Solomon's Ring*. Methuen.

MANNING, A. (1972).  *An Introduction to Animal Behaviour*, 2nd ed. Edward Arnold.

TINBERGEN, N. (1953).  *Social Behaviour in Animals*. Methuen.

TINBERGEN, N. (1969).  *The Study of Instinct*. Oxford University Press.

TINBERGEN, N. (1972).  *The Animal in its World. Field Studies*. Allen & Unwin.


*Parasites:*

CROLL, N. A. (1966).  *Ecology of Parasites*. Heinemann.

LAPAGE, G. (1951).  *Parasitic Animals*. Cambridge University Press.

ROTHSCHILD, M. and CLAY, T. (1961).  *Fleas, Flukes and Cuckoos*. Collins (New Naturalist series).

WILSON, R. A. (1967).  *An Introduction to Parasitology* (Studies in Biology No 4). Edward Arnold.


*Pollution:*

CARTHY, J. D., and ARTHUR, D. R., Eds., (1968).  *The Biological Effects of Oil Pollution on Littoral Communities* (Symposium Proceedings). Field Studies Council.

EDWARDS, R. W. (1972).  *Pollution* (Oxford Biology Reader No. 31). Oxford University Press.

GOODMAN, G. T., EDWARDS, R. W., and LAMBERT, J. M., Eds. (1965).  *Ecology and the Industrial Society*. Blackwell.

HYNES, H. B. N. (1960).  *The Biology of Polluted Waters*. Liverpool University Press.

LAVARONI, C. W. and O'DONNELL, P. A. (1971a).  *Water Pollution*. Addison-Wesley.

LAVARONI, C. W. and O'DONNELL, P. A. (1971b).  *Air Pollution*. Addison-Wesley

LAVARONI, C. W. and O'DONNELL, P. A. (1971c).  *Noise Pollution*. Addison-Wesley.

MELLANBY, K. (1967).  *Pesticides and Pollution*. Collins (New Naturalist series) and Fontana Paperback.

MELLANBY, K. (1972).  *The Biology of Pollution* (Studies in Biology No. 38). Edward Arnold.

SMITH, J. E., Ed. (1968).  *'Torrey Canyon': Pollution and Marine Life*. Cambridge University Press.

STOKER, H. S. and SEAGER, S. L. (1972).  *Environmental Chemistry: Air and Water Pollution*. Scott Foresman & Co.

WARREN, C. E. (1971).  *Biology and Water Pollution Control*. Saunders.

*Autecology:*

There are many sources of information about individual species. Even if there is no literature devoted to the plant or animal you wish to study, an examination of an account concerning a related species will give a useful indication of what to aim for. Some of the more useful sources are listed below; others may be traced through your local reference library.

*Biological Flora of the British Isles* – Consists of separate ecological accounts of individual flowering plant species, reprinted from *The Journal of Ecology* (Oxford: Blackwell).

Collins' New Naturalist Series Monographs – volumes devoted mainly to single species.

Forestry Commission Leaflets and Forest Records, including Nos. 34 *The Badger*, 44 *Voles and Field Mice*, 64 *Pine Martens*, 76 *Polecats* and 77 *Hedgehogs*.

Sunday Times '*Animals of Britain*' series, edited by L. Harrison Matthews (1962). Covers practically every British Mammal.

*Other useful sources are:*

BRISTOWE, W. S. (1958). *The World of Spiders.* Collins (New Naturalist series).

BURROWS, R. (1968). *Wild Fox.* David & Charles.

BURTON, M. (1968). *Wild Animals of the British Isles.* Warne.

CORBET, G. B. and SOUTHERN, H. N. Eds. (1977). *The Handbook of British Mammals,* 2nd edition. Blackwell.

CROWCROFT, P. (1957). *The Life of the Shrew.* Max Reinhardt.

EDWARDS, C. A. and LOFTY, J. R. (1972). *The Biology of Earthworms.* Chapman & Hall.

FROST, W. E. and BROWN, M. E. (1967). *The Trout.* Collins (New Naturalist series).

GODFREY, G. and CROWCROFT, P. (1960). *The Life of the Mole.* London Museum Publication.

HERTER, K. (1965). *Hedgehogs.* Phoenix House.

IMMS, A. D. (1971). *Insect Natural History* (New Naturalist Series). Collins.

LACK, D. (1953). *The Life of the Robin.* Penguin (Pelican Series).

MELLANBY, K. (1971). *The Mole.* Collins (New Naturalist series).

MURTON, R. K. (1965). *The Wood Pigeon.* Collins (New Naturalist series).

NEAL, E. (1969). *The Badger.* Collins (New Naturalist series).

NEAL, E. (1977). *Badgers.* Blandford Press.

SHORTEN, M. (1954). *Squirrels.* Collins (New Naturalist series).

SUTTON, S. L. (1972). *Woodlice.* Ginn.

THOMPSON, H. V. and WORDEN, A. N. (1956). *The Rabbit.* Collins (New Naturalist series).

TINBERGEN, N. (1953). *The Herring Gull's World.* Collins (New Naturalist series).

TWIGG, G. I. (1975). *Techniques in Mammalogy* (*Finding, Catching, Marking*). Mammal Review, 5, No. 3.

# Index